8.16.77

PERGAMON INTERNATIONAL LIBRARY
of Science, Technology, Engineering and Social Studies
*The 1000-volume original paperback library in aid of education,
industrial training and the enjoyment of leisure*
Publisher: Robert Maxwell, M.C.

Fuels, Furnaces and Refractories

International Series on
MATERIALS SCIENCE AND TECHNOLOGY
Volume 21—Editor: R. W. Douglas, D.Sc.

Other Titles in the International Series on
MATERIALS SCIENCE AND TECHNOLOGY

KUBASCHEWSKI, EVANS & ALCOCK
Metallurgical Thermochemistry, 4th Edition
RAYNOR
The Physical Metallurgy of Magnesium and its Alloys
PEARSON
A Handbook of Lattice Spacings and Structures of Metals and Alloys—
Volume 2
MOORE
The Friction and Lubrication of Elastomers
WATERHOUSE
Fretting Corrosion
DAVIES
Calculations in Furnace Technology
REID
Deformation Geometry for Materials Scientists
BLAKELY
Introduction to the Properties of Crystal Surfaces
GRAY & MULLER
Engineering Calculations in Radiative Heat Transfer
MOORE
Principles and Applications of Tribology
GABE
Principles of Metal Surface Treatment and Protection
GILCHRIST
Extraction Metallurgy
SMALLMAN & ASHBEE
Modern Metallography
CHRISTIAN
The Theory of Transformations in Metals and Alloys, Part 1, 2nd Edition
HULL
Introduction to Dislocations, 2nd Edition
SCULLY
Fundamentals of Corrosion, 2nd Edition
SARKAR
Wear of Metals
HEARN
Mechanics of Materials
BISWAS & DAVENPORT
Extractive Metallurgy of Copper

Fuels,
Furnaces and Refractories

by

J. D. GILCHRIST

B.Sc., Ph.D., A.R.C.S.T., F.I.M.

Department of Metallurgy and Engineering Materials,
The University of Newcastle-upon-Tyne, England

PERGAMON PRESS

OXFORD · NEW YORK · TORONTO
SYDNEY · PARIS · FRANKFURT

U.K.	Pergamon Press Ltd., Headington Hill Hall, Oxford OX3 0BW, England
U.S.A.	Pergamon Press Inc., Maxwell House, Fairview Park, Elmsford, New York 10523, U.S.A.
CANADA	Pergamon of Canada Ltd., 75 The East Mall, Toronto, Ontario, Canada
AUSTRALIA	Pergamon Press (Aust.) Pty. Ltd., 19a Boundary Street, Rushcutters Bay, N.S.W. 2011, Australia
FRANCE	Pergamon Press SARL, 24 rue des Ecoles, 75240 Paris, Cedex 05, France
WEST GERMANY	Pergamon Press GmbH, 6242 Kronberg-Taunus, Pferdstrasse 1, Frankfurt-am-Main, West Germany

Copyright © 1977 J. D. Gilchrist

First edition 1977

Library of Congress Cataloging in Publication Data

Gilchrist, James Duncan.
Fuels, furnaces and refractories.

(International series on materials science and technology; v. 21) (Pergamon international library)
A combined revision of the author's Fuels and refractories and Furnaces.
Bibliography: p.
1. Furnaces. 2. Refractory materials. 3. Fuel.
I. Gilchrist, James Duncan. Fuels and refractories.
II. Gilchrist, James Duncan. Furnaces. III. Title
TJ320.G49 1976   621.4'025   76–5865
ISBN 0–08–020430–9
ISBN 0–08–020429–5 pbk.

Printed in Great Britain by Biddles Ltd., Guildford, Surrey

Contents

Preface   vii

Introduction   ix

Part I.   Fuels

1. *Classification of Fuels*   3
2. *Properties and Tests*   6
3. *Coal*   17
4. *Carbonization*   29
5. *Coke*   37
6. *Gaseous Fuels*   45
7. *Liquid Fuels*   60
8. *Electrical Energy*   74
9. *Trends in Fuel Utilization*   90

1975042

Part II.   Furnaces

10. *The Evolution of Heat*   105
11. *The Combustion of Fuel*   121
12. *The Conversion of Electrical Energy to Heat*   139
13. *Heat Transfer*   150
14. *Thermal Efficiency*   172
15. *Furnace Aerodynamics*   183
16. *Furnace Construction*   193
17. *Classification of Furnaces*   197
18. *Laboratory Furnaces*   231

Part III.   Refractories

19. *Classification of Refractories*   237
20. *Properties and Testing*   240
21. *Manufacture of Refractories*   253
22. *Alumino–Silicate Refractories*   258
23. *Silica Bricks*   273
24. *Magnesite–Chromite Refractories*   284
25. *Carbon*   297
26. *Insulating Refractories*   300
27. *Special Refractories*   303

Part IV. Instrumentation

28.   *Pyrometry and Control*   311

Appendix   342

Bibliography   344

Index   347

Preface

WHEN the time came to prepare second editions of *Fuels and Refractories* and of *Furnaces* it seemed appropriate that the opportunity should be taken to combine them into a single volume. They are essentially about the same subject—fuel technology—and most students will surely be interested in that subject as a whole rather than about its constituent parts.

In the thirteen years since the first editions were prepared, important changes have been seen in the pattern of use of energy. These have been due to economic and political factors principally concerned with price differentials arising between coal and oil—the products of labour and capital intensive industries respectively. While oil was relatively cheap during the 1960s users changed over to its use and coal generally went into recession even in an expanding energy market. In 1972, however, the Organization of Petroleum Exporting Countries (OPEC) took political action which has resulted in the elimination and, at least for a time, the reversal of the differential. At the same time cost of coal has been raised by a succession of successful wage claims by miners in Britain so that the cost of energy generally has risen significantly relative to other commodities. This has brought about an improved appreciation among laymen of the fact that fossil fuels are not inexhaustible and a better understanding of the need for energy to be used economically and efficiently.

Compared with the political and economic changes of the period, technological changes in the field of fuel utilization have been small except in so far as industry has been adapting its practices to meet the changing economic circumstances. The general scientific principles underlying fuel utilization have not

vii

changed at all. Consequently many pages of this book will not be found to have been altered from the original editions. Among improvements introduced there is a selection of worked examples offered and SI units have been introduced though not exclusively in recognition of the fact that other systems are still used widely throughout the world—and some c.g.s. units like the calorie may yet survive on merit. Energy industries have been slow and some Imperial units are retained in the text particularly in association with some standard testing procedures. A table of conversion factors is included in an Appendix.

Introduction

IT is the purpose of this book to present a concise account of the sources of energy available to modern industry and to discuss how this energy can most efficiently be used.

Fuels are the raw materials consumed by furnaces, the main source of energy which has to be converted into heat and put to useful purposes raising temperatures and developing power. The amount of chemical energy available in any load of fuel is, of course, limited and is available on one occasion only, so it is desirable that it should be put to the very best use when that single opportunity is being exploited. Wasted fuel is a waste of natural resources. It is also a waste of money—or of human effort.

Electricity is not strictly a fuel but like the chemical energy of fuel it can be converted into heat and used for raising furnace burden temperatures as well as for reconversion into mechanical energy in a wide variety of equipment. It is the most versatile form of energy and the most easily distributed. The generation and distribution of electricity from both chemical and nuclear energy sources is discussed in association with fuels while its application to heating is included in the section on furnaces.

A furnace is a device in which energy is released as heat which is then used to raise the temperature of materials within, called the "burden". Furnaces operating at low temperatures up to about 300°C are called stoves or ovens, depending on their purpose, while others used at higher temperatures particularly those associated with the ceramics industries, are called kilns. Glass is melted in "tanks" and other special terms occur often suggestive of the purpose such as "sintering machine", used in metallurgical ore dressing. Other types are named according to purpose (annealing furnace), shape (shaft furnace) or function (boiler). In general the term "furnace" will be used to embrace all of these types until distinctions can be made in Chapter 17.

In the working of a furnace there are two major processes. These are the conversion of chemical energy into heat and the transfer of that heat into the burden. For reasons of economy these processes ought to be carried out with the maximum thermal efficiency and as fast as possible—or in accordance with the optimum compromise between these—and it is one of the principal aims of this book to consider how this can best be done. There is of course no general formula to be adopted but a large number of factors to be considered simultaneously, each carrying different weight under different circumstances. Consideration is given to the thermodynamics, physics, chemistry and kinetics of combustion of solid, liquid and gaseous fuels; to burner design, heat transfer and flow of gases through furnaces and flues; and to means of controlling energy supply rates and temperatures.

Frequently burdens have to undergo physical or chemical changes within a furnace, in which case it may become a high temperature reaction chamber. Its shape may then be determined by the nature of these changes or by the need for operations such as mixing to be carried out at temperature. When the burden is to be melted a hearth or crucible must be provided of suitable shape to contain the liquid formed and means must be made available for its discharge. Furnace design must take care of such matters.

Furnaces are built of refractory materials—heat resisting substances often in the form of bricks of standard shapes made from selected clays and other rocks blended, moulded to shape and fired at high temperature. These bricks are assembled within a steel or cast iron framework or shell to form the furnace, flues and other accessory parts which reach high temperatures. In many uses the brickwork is very durable and features as an almost permanent capital asset, but deterioration with time is more normal and in extreme cases, particularly though not exclusively in the metallurgical industries, conditions can be so severe that the useful lives of refractory linings or roofs may be only a few weeks and indeed sometimes only a few hours. In these cases refractories are consumables and any better choice of brickwork or amelioration of conditions which will increase refractory life is of great value, not only because the cost of the refractories themselves is reduced but also because lost production time required for rebuilding is also reduced. The intelligent use of refractories, like the efficient use of fuel can save a lot of money. It is customary to consider the cost of refractories per ton of product so that considerable increases in the price per brick can be absorbed if the more expensive brick lasts longer in the furnace, production rates being unchanged, or if it allows a higher production rate to be attained even if the brick life in hours is not increased.

There is a bias in the book toward the metallurgical uses of fuel which may be criticized by the reader whose interests lie elsewhere but the bias is not inappropriate in so far as the metallurgical industries have a wider range of experience than any other in fuel technology and most of the modern high quality refractories have been developed for or by the steel industry. The metallurgical industries are major users of fuel and they use all kinds of fuels in the widest range of types of furnace including the largest built; they develop the highest temperatures and operate under some of the most arduous conditions chemically and mechanically; their burning rates and flame intensities are among the highest used; and they can claim a large share of the credit for the advances in fuel technology which have taken place since the industrial revolution. These advances have included such "classical breakthroughs" as the invention of the hot blast stove in association with blast furnaces and the development of the regenerative principle upon which the open hearth steelmaking process depended. In more recent times the steel industry has done much to develop the use of oxygen (rather than air) for the combustion of oil and gas—a phase of development which is still not complete. Continuing demand for higher temperatures and faster production rates ensure that there is still plenty of scope for research and development and for further improvement into combustion techniques.

Other industries, of course, make their own contributions to the improvement of fuel utilization, not least the electricity industry whose most recent conventional generating plant operates at efficiencies undreamt of forty years ago. The choice and availability of fuels have improved in recent years particularly for the domestic consumer while the design of domestic heating equipment has undergone a major revolution. These improvements have the double advantage that they help the consumer to meet the requirements of the Clean Air Act while at the same time by their greater efficiency they reduce the cost of achieving any required level of comfort.

No discussion about fuel can ignore economics. Fuel is used in almost every industry and its cost is often an important factor in determining the price of the product. A country which enjoys a plentiful supply of cheap indigenous energy should be at a considerable advantage when selling its manufactured goods over countries not so favoured and indeed the established industrial areas of the world are with few exceptions but notably excepting Japan, also major coal producing areas.

The fuel bill of Great Britain in 1975 is probably about £7000 million per annum or about 10 per cent of the national turnover. Obviously it is in the interests of both the individual and the nation to get the best possible value for this money. The individual may exploit the current advantages in price or in technology to be obtained by using one fuel rather than another at any particular time and governments not infrequently influence individuals in that choice by the application of taxes or subsidies. Ultimately, however, value for money is best assured by organizing for the highest possible efficiency of utilization of fuel at all points in the economy—not only in steelworks and power stations but in the shops, offices and homes, for homes use more energy than steelworks and it is there that the potential savings are still greatest.

Unfortunately it is usually necessary to make a substantial capital outlay in order to convert to efficient equipment and it is not always apparent that money will in fact be saved in the change over. There is probably a good case for subsidising such conver- . sions. There is certainly a case for prohibiting the installation of new equipment which does not utilize its fuel with an efficiency which is well up to the best that is currently possible.

PART ONE

Fuels

1

Classification of Fuels

OMITTING consideration of electrical energy until Chapter 8, fuels can be classified in several different ways.

Most fuels are "fossil fuels", that is coal, or oil, or their derivatives. The few exceptions are important only for very restricted purposes. They include, for example, metallic aluminium in the "thermit" process, and sulphur in some roasting operations There are many metal extraction processes where the heat of reactions contributes a high proportion of the energy requirements as in the conversion of iron to steel by the Bessemer or other pneumatic processes, where the fuel may be said to be silicon, carbon and phosphorus. Another non-fossil fuel is wood with its derivative, charcoal. Peat and lignite are included among the coals but peat especially is on the borderline between "vegetable" and "fossil". Wood, charcoal and peat are not used in large quantities by industry today but they are not obsolete. Reserves of peat are large and it is used in at least one place for generating electricity. Charcoal too is still used in small iron blast-furnaces in countries where wood is more plentiful than coal, or where special metallurgical requirements have to be satisfied. Vegetable waste like spent sugar cane, and sewage gas provide useful local supplies of energy in undeveloped and in highly industrial regions respectively.

Fuels may be classified as raw, or prepared. Thus we have raw coal and crude oil on the one hand, and on the other, prepared or refined products like coke, refined oils and pitches. The proportion of raw fuel used is decreasing. Coal would seldom be used unwashed and ungraded, and burning appliances are now designed to accept particular sizes of coal or coke. Cokes and prepared fuels like "Coalite" are becoming more necessary as atmospheric pollution is more and more discouraged, but again good design in burning appliances can extend the use of raw coal. Metallurgically, coke is the most important derivative of coal, but several kinds of gas are also prepared and used in vast quantities.

Oil is seldom used crude, but is usually refined by fractional distillation and other processes, different fractions being put to different purposes. The lightest fraction of all—natural gas—is often available at the oilfield in very large quantities. Where circumstances permit, this gas is piped to industrial areas. It can also be liquefied and transported in tankers over great distances. It is the heavy fractions which are most often used in metallurgical industries, often so heavy that they have to be warmed to permit pumping through pipes. Even the residual pitch is available as a fuel.

Fuels may be solid, liquid or gaseous, as already implied. Liquid and gaseous fuels can more conveniently be handled in pipelines than solid fuels. They can be more or less on tap, whereas coal and coke must usually be moved in batches. To this there are exceptions, however. Lump coal can be taken long distances within a plant on conveyor belts and can be fed to furnaces on moving grates and pulverized coal can be blown with air along pipes. On the other hand, oil, while readily handled in hydraulic systems, is usually delivered by tanker in batches, and pipeline distribution is as yet only between major oil storage installations. Solid fuel is most readily stored, in the open if necessary, though not without some difficulties being encountered. Oil and gas must be held in closed containers, each of limited, if large capacity and the bleeding to waste of unwanted excess by-product gas is sometimes an unfortunate necessity. The use of old gas wells or similar suitable strata for storing gas and possibly oil too has been suggested and might afford a useful buffer against seasonal variation in demand.

Fuels are sometimes referred to as being rich or lean. This refers to the available calories per unit of mass or volume. Coke with 90 per cent of carbon is richer than say peat which has a high water content. Natural gas is richer than say producer gas which has a high nitrogen content.

Fuels may also be distinguished as being replenishable or irreplenishable. The fossil fuels on which we rely today all fall into the latter category nor are the radioactive elements from which nuclear energy can be obtained available in unlimited quantities unless perhaps means can be found to extract them from the oceans. Wood and peat can be regenerated but at rather low rates. Water, wind and tides will continue to provide energy indefinitely but at a restricted rate which will probably never come near to matching our needs. Solar and geothermal energy are inexhaustable but are not yet being tapped on a large scale. Only the dream that the energy of the fusion of plentiful light elements may one day be controlled gives hope that there may be ample energy for ever.

2

Properties and Tests

THESE can be dealt with here only in a general way since different fuels have peculiar properties and special tests which can best be discussed separately at later stages. The flashpoint of an oil, for example, has no counterpart in solid fuels but all fuels have a calorific value and a chemical composition and it is this latter type of property which will be discussed here.

Chemical Composition

Fossil fuels and their products are composed of mixtures of organic compounds of carbon, hydrogen, oxygen, nitrogen, sulphur, etc., along with some inorganic matter which is identified in the residual ash after combustion.

The carbon and hydrogen are of greatest importance as heat producers, other elements being at best diluents. Oxygen, for example, would usually be combined with hydrogen or carbon in compounds which would have to be endothermally dissociated during combustion, so reducing the effective heating effect of the carbon and hydrogen.

The organic compounds in coal are very complex, usually called humic acids and with molecular weights of several thousands. The oils contain a wide range of hydro-carbons—paraffins, olefines, naphthalenes and aromatics—ranging from the simplest compounds like methane (CH_4) up to waxes of very large molecular weight. Normally these are separated according to volatility before being marketed, each fraction for a particular purpose.

Sulphur is an important impurity in both coal and oil fuels. In coal it is partly mineral and partly organic, being present at about 1 per cent in British coals, but it may be much higher in some seams and rises to over 6 per cent in some countries including the USA. The mineral part is largely FeS_2 which is visible in seams and veins but too finely disseminated to be economically separated. In oil it is wholly organic and in this form it cannot possibly be eliminated completely from the fuel. The proportion of sulphur in oil is also about 1 per cent, but may be lower, or higher up to about 3 per cent, when the fuel becomes metallurgically unsuitable. Low sulphur fuel is in great demand as it reduces the amount of metallurgical work necessary particularly in steelmaking. It is therefore relatively costly.

Nitrogen is present in coal at between 1 and 2 per cent and in crude oil at under 0·5 per cent. Oxygen also is very low in oil but in coal it may rise to about 20 per cent and even higher in lignite, being an essential part of the coal substance.

The free moisture content particularly of solid fuels is important to know, when buying, when using or when analysing. Coals and cokes have a large internal area due to their finely porous structure and therefore adsorb moisture from the atmosphere. This leads to an inherent moisture content dependent on the ambient humidity. A much higher value is frequently encountered due to rain, washing (of coal) or quenching (of coke), and it should be kept to a minimum by storage and handling practice.

Inorganic matter, determined as ash after total combustion of coal in air, is probably mainly mineral in origin, though some is very closely associated with the coal substance or so finely disseminated as to be physically inseparable from it. A part of the ash will probably come from "dirt" bands and from the shale bands overlying and underlying the coal seam, but most of this should be separated by washing. The ash content of coals may be between about 1 and 15 per cent, but the range 5–8 per cent is probably typical for British coals. The ash is mainly silica, alumina and ferric oxide with varying amounts of other oxides such as CaO,

MgO and Na_2O. This is not, of course, the condition of the inorganic matter in the coal, and the decomposition and conversion of clays, shales, carbonates and sulphides into oxides or even into alumino-silicates, involves some change in weight, usually a loss. In coke, on the other hand, the original minerals have already been decomposed and probably reduced, and ash may be heavier than the inorganic content of the fuel due to the oxidation, particularly, of iron.

The ash content of heavy oil is only about 1 per cent, and it is less in light oils and greater in pitch, in which it concentrates in the fractionation processes. This ash is rather different from coal ash in that it may be rich in sodium sulphate and vanadate, the latter derived from the remains of fossilized fish. This is of consequence when the high-temperature corrosive properties of these substances are likely to be effective, especially in internal combustion engines and gas turbines.

The fusion point of ash is also important and varies from about 1050°–1500°C. In Britain more than half the seams yield coals whose ash fusion point exceeds 1200°C and these include the best coking coals. Of the remainder only a very few have fusion points below 1100°C. This fusion temperature determines the form in which the ash is most likely to be removed from the furnace—as a dry ash, as a clinker, or even as a slag—and this obviously has an important bearing on details of furnace design. Generally, if the fusion point exceeds 1200°C clinkering may occur but does not often give trouble. Ash is, however, likely to be non-homogeneous and clinker can be formed by the cementing together of quite refractory pieces of ash with relatively small amounts of fusible material, particularly if high in iron oxide. Much of the ash of pulverized coal is carried off as "fly-ash" in the chimney gases, but if it is fusible it can gradually build up on the walls of the furnace chamber or in the flues. Fly-ash, if produced in large quantities, as for example by large electricity generating plants, may have to be precipitated out of the chimney gases before discharge to avoid offence to the public.

The fuel analysis determines the volume of the air required for combustion and the volume and composition of the combustion gases. This information is required for the calculation of flame temperatures and when burners and flues are being designed. If the necessary air cannot be supplied or if the products of combustion cannot be removed an adequate supply of heat cannot be maintained.

Example. Calculate the volume of air required and the volume and composition of the flue gases when 1 kg of oil is burned stoichiometrically if the oil analysis is 87 per cent C, 13 per cent H. From the reaction

$$C + O_2 = CO_2$$

it can be seen that 12 g carbon requires 1 g molecular volume or 22·4 l. of oxygen and produces 22·4 l. CO_2 at S.T.P. Similarly from

$$2H_2 + O_2 = 2H_2O$$

it can be seen that 4 g H_2 requires 22·4 l. of oxygen and produces 44·8 l. H_2O vapour at S.T.P. Therefore 1 kg oil requires

$$(870/12 + 130/4) \times 22·4 \text{ l. } O_2$$

i.e. 2350 l. O_2 or 2·350 m³ O_2

In air this is associated with 2·35 × 79/21 or 8·32 m³ N_2. So that the air required to burn 1 kg of oil equals 10·67 m³. The combustion gases consist of 8·32 m³ N_2,

$$870/12 \times 0·0224 = 1·623 \text{ m}^3 \text{ } CO_2$$

and $$130/4 \times 0·0448 = 1·458 \text{ m}^3 \text{ } H_2O.$$

Therefore the volume of the combustion gases from 1 kg of oil is

$$11·40 \text{ m}^3 \text{ at S.T.P.}$$

The flue gas analysis by volume is

73 per cent N_2, 14·2 per cent CO_2 and 12·8 per cent H_2O.

Any oxygen already present in the fuel is available toward these reactions. The remaining oxygen requirement must be supplied, normally as air with a composition 21 per cent O_2 : 79 per cent N_2 by volume or 23·2 per cent : 76·8 per cent by weight. If oxygen enrichment is employed these natural ratios must of course be modified.

The S.T.P. volumes can be adjusted to true volumes if the pressure and temperature are known. In practical situations the pressures are usually close to atmospheric while the temperatures are initially very high but fall rapidly as the gases pass through the system. In practical situations too, a small excess of air is usually admitted to ensure that combustion is rapidly completed, so that rather more nitrogen and some free oxygen would normally be found in the flue gases.

It is sometimes useful to compare fuels by the volume of flue gases produced per unit of heat energy produced. Obviously leaner fuels require greater flue capacity to vent the combustion products unless there are compensatory factors like oxygen enrichment of the air or a high level of preheat.

Gas analyses are usually in terms of volume percentages of the various components—H_2, CO, CH_4, CO_2, N_2, etc. In this case the oxygen and hence the air requirement for complete combustion is calculated in a similar manner:

$$2CO + O_2 = 2CO_2$$
$$CH_4 + 2O_2 = CO_2 + 2H_2O$$

Two volumes of CO require one of oxygen but one of methane requires two of oxygen, and so on.

The ultimate analyses of *solid fuels*, that is for all elements present, are determined by methods generally adopted in organic chemistry. Carbon and hydrogen are determined together by combustion to CO_2 and H_2O which are absorbed and weighed. Nitrogen is converted to ammonium salts under sulphuric acid with a selenium catalyst in the Kjeldahl method, the ammonia then being distilled off and determined by colorimetric or volumetric means. Sulphur is determined by combustion to sulphate either in an oxidizing fusion or by O_2 in a bomb, converted to $BaSO_4$ and weighed. Ash analysis is conducted separately by standard inorganic methods (as for ores and slags). Care has to be taken to make due allowance for moisture content (at the time of weighing out the sample) and the distinction has often to be made between organic sulphur and that part which appears in the ash.

For day-to-day control "proximate" analysis is usually sufficient. In coal this involves standard tests for moisture, volatile matter, fixed carbon, ash and sulphur. The volatile matter is determined by heating a dry sample out of contact with air at 925°C. Loss of weight is "Volatile Matter" (V.M.) and the residual weight is "Fixed Carbon" + "Ash". Ash alone is determined on a separate sample by total ignition in air at 800°C and hence fixed carbon is calculated as the difference. The fusion point of the ash, if required, is determined by comparison with Seger cones as in the refractories test on bricks.

Oil fuels undergo a variety of analytical tests, which vary with the kind of oil and the uses to which it is to be put.* *Specific gravity* may be required for conversions of volumes to weights or to meet specifications and would be of interest when oils were being mixed or blended. *Viscosity* is of interest when heavy oils have to be pumped through supply lines and it may be necessary to have data over a range of temperatures. *Flash point* is an "ignitability test" and provides information about handling dangers on the one hand, and ease of lighting up on the other. *Distillation tests* indicate what fraction or blend of the oil one is handling. Taken together these tests go far to define the "type" of oil, in terms light, medium and heavy.

Moisture may be determined in oils by settling out or centrifuging if in large amounts or by a standard distillation method. It is most likely to be present in the heavy fractions. *Sulphur* is determined by burning the oil sample in special apparatus in which the products of combustion can be drawn through an absorbent

[1]The Institute of Petroleum *Standards for Petroleum and its Products*, Vol. 1.

solution (Na_2CO_3) and the acidity due to SO_3 measured volumetrically, or, in the case of heavy oils, by oxidation with oxygen in a bomb and determination as sulphate. *Ash* is determined by evaporation of most of the oil followed by ignition of carbonaceous residues. Other special purpose tests distinguish aromatics from paraffins and measure carbon residues and asphaltenes which are undesirable in engine fuels but of little consequence in furnace fuels.

Gaseous fuels are analyzed by molecular species present, the results being expressed in percentages by volume. A convenient volume of gas is isolated (at a temperature and pressure which must be kept constant). It is then brought into contact with a series of reagents which absorb the constituents one at a time, the volume lost at each stage being noted. CO_2 is absorbed by KOH; O_2 by alkaline pyrogallol; and CO by ammoniacal cuprous chloride. Unsaturated hydro-carbons can be absorbed by fuming H_2SO_4. Hydrogen and methane and other saturated hydrocarbons cannot be absorbed in this way, and must be oxidized to H_2O (which condenses) and CO_2. There are several ways of doing this. Explosion with an excess of oxygen is a common method but slow combustion methods have some advantages. Oxidation by a heated CuO spiral converts only CO and H_2 to CO_2 and H_2O, leaving CH_4, C_2H_6, etc., unaffected. This can be followed by combustion on a platinum (catalyst) spiral when the quantities of CO_2 and H_2O formed will indicate the proportions of CH_4 and C_2H_6 in the gas. Obviously an accurate gas analysis is not easy, and complete separation not always possible in a single series of operations. SO_2, H_2S and HCN go with CO_2 while C_2H_2 will be absorbed along with CO unless special extra steps are incorporated. (See also page 329.)

Calorific Value

Chemical composition determines the calorific value (c.v.) of a fuel. This is the amount of heat obtainable by complete combustion of unit quantity of fuel. Unfortunately, it can be expressed in several different ways and it is often necessary to convert from one system of units to another.

The unit of heat is, for scientific purposes, most conveniently the *calorie*, or gramme calorie, being the amount of heat required to raise the temperature of 1 g of water from 15° to 16°C. The British Thermal Unit (B.t.u.) is frequently used in industry, and is the amount of heat required to raise the temperature of 1 lb of water from 60° to 61°F. There is a further Centigrade Heat Unit (C.H.U.) which compromises between the other two and is the heat required to raise the temperature of 1 lb of water by 1°C.

The unit quantity of fuel should then be the gramme or the pound but gaseous fuels are usually measured by volume so we get B.t.u. per lb; B.t.u. per cubic foot; C.H.U. per cubic foot; calories per gramme, and so on.

The Imperial system of units has survived in industrial and commercial use not only in Britain but also in the U.S.A. for longer than might have been expected but for scientific purposes the metric c.g.s. system has been in general favour for many decades. Legislation is now well advanced in Britain, however, to impose the Systeme International d'Unites—S.I.—on British industry and commerce—and on the scientific community also. This is like the c.g.s. system, a metric system which it was originally hoped would be completely self-consistent but so many of its units are inconvenient and unrealistic that concessions are still being sought and reluctantly granted and it may still be hoped that many of the advantages of the more rational c.g.s. system may yet be retained. There is little sign that the rest of the world is hastening to fall in line with Britain in this matter and it would not be surprising if like Esperanto it were to fizzle out.

In the S.I. the unit of energy is the Joule (J) which is the energy expended when a force of 1 Newton (N) is exerted through a distance of 1 metre (m)—a Newton being the force required to produce an acceleration of 1 ms^{-2} on a mass of 1 kilogramme (kg). Applied to the calorific value of a fuel this definition has no apparent relevance and it becomes necessary to accept as a fact that 1 Joule = 0·238846 calories (or 1 calorie = 4·1868 Joules). Calorific values in the S.I. have to be expressed in kJ/kg which is the same numerically as J/g, and is equal to 0·239 cal/g; or in kJ/m³, numerically the same as J/l and equal to 0·239 kcal/m³ or 0·239 cal/l.

The Joule can be considered to be the amount of heat required to raise the temperature of 0·2389 g of water from 15°C to 16°C. This is not, of course, a definition of the Joule but a measure of its size which it is necessary to know if it is to be used in thermo-chemical calculations.

Here, c.g.s. energy units will be used along with S.I. units. Imperial units will be omitted except in Appendix "A" where a selection of conversion factors is offered. The need to convert units will be with us for a long time even if it is only to make the literature of yesterday intelligible.

A distinction must be drawn between *Gross* and *Net Calorific Value*. When hydrogen is present in a fuel, water vapour is one of the products of combustion, and if this is condensed the latent heat liberated may be added to the other heats of combustion, to give the Gross Calorific Value of the fuel. Indeed the temperature to which the condenser in the test equipment is cooled is specified at 60°F (in Britain). In many practical cases, however, the combustion gases are not condensed in the working part of a furnace and the Net Calorific Value of the fuel, i.e. omitting the part available from water vapour between 100° and 15°C, is of greater interest. Obviously the difference is often very small and quite unimportant (unless costing is being based upon it), but in those cases where hydrogen or methane is a major constituent the difference is large and it is important to use the value appropriate to the purpose in hand.

In principle, calorific values can be calculated from complete analyses, if all the heats of reactions occurring during combustion are known. Certain assumptions have to be made, notably that combustion is complete to CO_2 and H_2O. This can be arranged in laboratory test equipment but often does not happen in a working furnace. Dissociation of compounds present in the fuel should also be taken into account and this is often very difficult as the compounds involved are many and complex. Usually formulae used to convert analyses into calorific values are empirical, and based on measurements on similar fuels using calorimeters.

Calculation of calorific value is most successful in the case of gaseous fuels where compounds are relatively simple and their net heats of combustion are usually known. Accuracy depends only on that of the gas analysis.

$$H_2 + \tfrac{1}{2}O_2 = H_2O \qquad \Delta H = -57,800 \text{ calories (net)}$$
$$= -242\cdot0 \text{ kJ}$$
$$CH_4 + 2O_2 = CO_2 + 2H_2O \qquad \Delta H = -191,800 \text{ calories (net)}$$
$$= -802\cdot8 \text{ kJ}$$
$$CO + \tfrac{1}{2}O_2 = CO_2 \qquad \Delta H = -67,623 \text{ calories}$$
$$= -283\cdot1 \text{ kJ}$$

To calculate the c.v. of a mixture, the analysis should be converted to gramme-molecular volumes, and the appropriate proportions of the gramme-molecular heats of the reaction added together and the sum related to unit value of gas.

Example. Calculate the net calorific value of a gas mixture containing 50 per cent H_2, 40 per cent CO, 5 per cent CH_4, and 5 per cent N_2.

Consider 1 litre of gas.

This contains $0\cdot5/22\cdot4$ moles H_2, $0\cdot4/22\cdot4$ moles CO, and $0\cdot05/22\cdot4$ moles CH_4. The heats of oxidation are $-57,800$, $-67,623$ and $-191,800$ cal/mole respectively. Therefore the total heat of oxidation of 1 litre of gas is

$$[(0\cdot5 \times 57,800) + (0\cdot4 \times 67,623) + (0\cdot05 \times 191,800)]$$
$$\div 22\cdot4 \text{ cal,}$$

i.e. 2920 cal/litre or 2920 kcal/m^3 or 12,200 kJ/m^3.

Example. Calculate the gross calorific value of the same gas.

Consider the reactions

$$2H_2 + O_2 = 2H_2O$$

and

$$CH_4 + 2O_2 = CO_2 + 2H_2O.$$

The volume of water produced in the vapour state is equal to the volume of hydrogen in the fuel gas plus twice the volume of methane,

i.e.   0·6 litres H_2O vapour per litre gas burned

or   0·6/22.4 moles H_2O per litre gas burned

or   18 × 0·6/22·4 g H_2O per litre gas burned.

For each gramme, (540 + 85) cal, i.e. 625 cal or 2610 J are released as steam at 100°C is condensed and cooled to 15°C.

Therefore the gross c.v. = 2920 + 625 × 18 × 0·6/22·4

= 3220 cal/litre

= 13,500 kJ/m³

Direct determination of c.v. by calorimeter is carried out in standardized equipment. Solids and liquids are probably best determined using a bomb calorimeter in which the sample is ignited in an excess of oxygen contained in a small pressure vessel immersed in a water bath insulated from its surroundings.

In modern bomb calorimeters the water bath is completely surrounded by a water jacket, the temperature of which is controlled electronically at exactly the same temperature as the water bath itself, following its temperature rise throughout the determination so that net heat loss from the water bath is impossible. The rise in temperature of the water bath is measured and the heat produced by the reaction calculated via the thermal capacity of the apparatus.

Gas calorimetry involves burning gas at a constant rate, and transferring the heat from the gases produced into a counter-flowing stream of water via a suitable heat exchanger.

Calorimetry usually gives gross c.v. but the Boys' gas calorimeter collects the condensed water for measurement to enable the conversion to be made accurately to net c.v. by a calculation similar to that shown above.

3

Coal

Origins

COAL is certainly derived from wood and other vegetable matter, decomposed first by bacteria under anaerobic conditions such as are still to be seen in some tropical swamps, and later modified by temperature and pressure.

Although coal-like substances can be produced from cellulose and resins in the laboratory, the exact conditions under which coal was formed are not known, nor are the variations in conditions which led to the variety of coals which are found. Factors which might be relevant are the degree of de-aeration in the early stages, the pH in the early stages, temperature and pressure in later stages and the time during which each stage persisted. The nature of the original vegetable matter may also have been important, particularly where segregation of parts like spores and cuticles by water flow may have occurred. Some coals have been modified by heat treatment due to their proximity to igneous intrusions.

Coals occur in seams which vary in thickness from a few inches to several feet. In the major coal-fields there are many such seams separated by thicker beds of shales, clays, sandstones and other sedimentary rocks. These series of strata may be faulted or folded, but across any coal-field the properties of the coal from any particular geological horizon are substantially constant while the quality varies more or less regularly from one seam to another in the vertical direction. For some purposes, notably carbonization, it is very desirable that the coal shall be drawn from a particular seam, and the exhaustion of such a seam would lead to temporary difficulties for the user until a new source of a similar coal could be found.

Coal can be won from seams down to 18 inches thick by conventional deep-mining methods if the quality is high enough and conditions favourable. Most coal is deep-mined but open-cast mining opportunities exist in most coal-fields and this can be the more economical process if the overburden does not exceed about thirty metres. Out-cropping seams may be altered by weathering near the surface but in general open-cast coal is similar in quality to deep-mined coal from the same seam.

Classification

There are a number of classification systems for coals—not all compatible with one another.

TABLE 1

Typical Analytical Data for the Range of Coals, and Coke

Fuel	Ultimate—dry ash free basis			Proximate—air dry basis			Air dry C.V. (net)	
	%C	%H	%O	% Moisture	% V.M.	% Ash	cal/g.	kJ/kg.
Peat	60	6	34	20	70/60	1/10	3500	14,650
Lignite	70	8	22	15	50/40	8/12	5000	20,900
Sub-bituminous coal	75/82	6/5	20/12	10	40/30	5/10	5500	23,000
Bituminous coal	82/90	6/4·5	12/3	2	35/20	5	7750	32,440
Semi-anthracites	91/93	4	4	1	10	5	8000	33,490
Anthracite	94	3	2	1	8	3	8000	33,490
Coke	95	1	2	2	8	7	7300	36,560

A rough preliminary grouping is given in Table 1. Peat and lignite are of no metallurgical importance, but where available in quantity they are dried and used industrially. Peat is used for the generation of electricity in Eire and lignites for gas production and domestic purposes, the largest user being Germany. Anthracite is a scarce and expensive high-grade fuel used industrially for steam raising and central heating. It is virtually smokeless, which makes it useful in malting. It is also used metallurgically as a carburizing agent and deoxidizing agent. The important range of coals is the bituminous group which merges into the anthracites at about 90 per cent carbon. These can be further classified in a manner which suggests the uses to which they might be put, as shown in Table 2.

TABLE 2

Typical Compositions of Some Bituminous Coals

	Carbon %	Hydrogen %	Oxygen %	V.M. %
Long flame, non-caking steam and house coals	83/86	5/6	6/12	30/40
Long flame, partly caking gas coals	82/86	4·5/5·5	5/9	30/40
Short flame coking coals	85/89	4·5/5·5	4/7·5	20/30

Industrially, this classification is of great importance, particularly in the separation of coking and non-coking types. Obviously classification by carbon content or "rank" is important in separating coals of very different characteristics. The use of volatile matter content, which is roughly complementary to rank, would divide up coals in a similar way. Within these divisions, however, there may be further sub-divisions and even an overlapping of these from one main division to another, so that the scheme is never so clear-cut as suggested above. Important factors other than carbon content include oxygen and hydrogen content and the ratio between them. This probably reflects the nature of the original organic matter and in turn affects the coking and caking properties. The Seyler classification attempts to cover all these

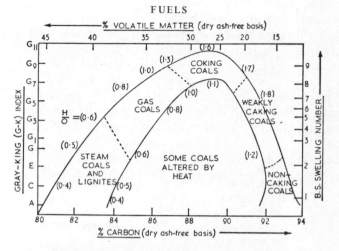

FIG. 1a. A summary of the relationships between rank, volatile matter content, hydrogen : oxygen ratio, and coking characteristics of coals. The numbers in brackets are the H/O ratios at the upper and lower edges of the band which contains most normal coals. At any given rank, in general, a high proportion of hydrogen to oxygen favours strong swelling and agglutination.

factors. A simpler diagrammatic representation is presented in Fig. 1a which broadly indicates how quality is related to composition. The Gray-King assay is explained on page 22.

For commercial purposes an even simpler classification is required and various systems have been devised by official bodies. In Britain the National Coal Board has designated each class of coal by a number as indicated in Fig. 1b. Each class lies in a particular range of volatile matter content *and* has its Gray-King Index within a particular narrow group. Classes are grouped in series—the 100 series, the 200 series and so on, the series being sub-divided into narrower ranges of volatile matter content. While series of low volatility are distinguished from each other by their volatile matter contents, the series of high volatility differ from each other in respect of their G–K indices only. The classes 206 and 305 are heat altered coals. The correspondence between Figs 1a and 1b will be obvious with coals ranging from non-caking, low volatile anthracites (100 series), through the prime coking coals (300 series) to the lignites (900 series). It should be noted that the volatile matter scale is linear in Fig. 1b but not in 1a, and also that in Fig. 1b the volatile matter content is not on the d.a.f. basis.

Four types of coal substance can be recognized visually. *Vitrain* is bright, black and brittle and usually in thin bands. It fractures conchoidally. It contains no obvious plant structure and seems to have been derived from bark. *Clarain* is not so bright and has an irregular fracture. It contains more plant remains (spores, etc.) than vitrain and is the commonest of the four types of coal substance. *Durain* is duller, greyer and very hard, and cracks irregularly. It is highly charged with durable plant remains (spores, cuticle, etc.), and it is thought that durain was formed from silts or muds of small particles of vegetable matter. Cannel coal may

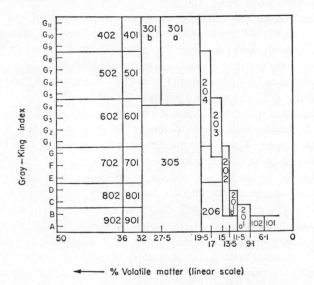

← % Volatile matter (linear scale)

FIG. 1b. National Coal Board commercial classification of coals based on volatile matter content and Gray–King assay.

be an extreme form of durain. It is associated with a high clay content and in extreme cases may become a sort of shale. *Fusain*, a soft powdery form occurring only in thin seams between bands of the other types, seems to have been formed differently from them. Massive coal fractures through fusain bands. About half of a coal seam may be clarain, a sixth to a third durain. Vitrain may rise to 10 or 15 per cent, while fusain is only present up to 1 or 2 per cent.

In coal preparation, durain and clarain may segregate as "hards" and "brights" respectively. While in any seam the analyses and characteristics of hards and brights are of the same order, there would usually be some small difference in properties, particularly in so far as the brights, vitrain and clarain, would have rather greater coking power.

A system of microconstituents in coal has been suggested, the important ones being the hydrogen-rich exinite, derived from spores and cuticles, fusinite which is relatively low in hydrogen, and vitrinite of intermediate hydrogen content. Clarain and durain are mixtures of these. Seyler found ten varieties of vitrinite whose reflectivities varied in geometric progression. In any coal only three or four of these are found, one predominating and this is determined by the rank of the coal.

Coals are sometimes examined by solvent extraction techniques which demonstrate the differences between coals but do not explain them. The swelling characteristics of coking coal can be destroyed by solvent extraction.

Heating to release gas and tar is embodied in other empirical tests of interest to the carbonizing industry, which is also interested in the quality of the coke residue. The best known test in this group is the Gray-King assay in which 20 g of sample is heated in a silica tube to 600°C and the coal visually classified A to G by comparison with standard photographs depending on whether the residue remains a powder (A), or coalesces into a hard coherent mass with the same volume as initially (G). Intermediate stage B is non-caking, C and D are weakly caking while E, F and G are medium caking. Strongly caking coals swell and are designated G_1, G_2, etc., to G_{10}, the suffix indicating the number of grammes of inert carbon which must be blended into the 20 g charge to give zero swelling. The G–K index correlates well with rank and volatile matter and is probably the best index of the caking capacity of a coal or coal blend. Another common test gives as an index the British Standard Swelling Number, 1–9. A sample of coal is heated at a standard rate in a crucible with a lid on until volatiles cease to be evolved and again the shape and size of the button are compared with standard charts. The relationships between the agglutinating properties of a coal or coal blend and rank, volatile matter content and hydrogen/oxygen ratio are summarized in Fig. 1. A simple diagram of this kind is rather approximate and ignores many exceptional cases and must not be used to predict the behaviour of any individual sample.

The swelling properties of coals are associated with their physical structures. On the sub-microscopic scale coal is composed of tiny crystallites called micelles. These have a layered structure like graphite but are much more complex chemically and they are only 30–40 Å (3–4 nm) in length or in diameter. Consequently there is a very high proportion of misfit volume in the structure and this can easily be demonstrated to amount to about 35 per cent of the total volume in bituminous coal, that is 35 per cent of sub-microscopic porosity pervades the structure. That figure will depend on the type of coal.

When coal is heated, gases are evolved, escaping from each tiny micelle into this finely distributed pore space. At the same time, at 300°C, the bonds between the layers in the micelles are disrupted and the layers slide over each other giving the coal the appearance of having melted. If the gases cannot find their way out at this stage they form bubbles which expand the material and press particle upon particle to produce a coalesced mass with the characteristics of a stable foam. At a rather higher temperature the residual micellar structure reforms with a new system of interlamellar bonds and the porous mass acquires a new rigidity—that of the char which may be hardened by further heating to form coke. Should the quantity of gas escaping be small (as from a non-caking coal, Fig. 1a) or should the permeability of the system of sub-microscopic pores be high enough to let a larger quantity of gas escape easily (steam coal) insufficient gas pressure will build up to effect coalescence and swelling. The optimum condition for swelling occurs where the volatile matter content is moderately high at 25–30 per cent but the pore system has too low a permeability to let this escape. More volatile coals have wider pores, while less volatile coals have less gas trying to escape. Any particular coal will exhibit a higher degree of swelling if heated up more rapidly because gas will be liberated more quickly without there being any increase in the permeability of the system. Coals may be blended in which case particles of strongly swelling coal will expand and envelop particles of non-caking coal so that the blend will appear to behave as a moderately swelling coal. Traditionally, coke was made from strongly swelling coals slowly heated in beehive ovens but now blends are used which when rapidly heated in vertical ovens produce coke with equally good properties. Diminishing reserves of strongly coking coals are thus conserved by dilution with coals of more frequent occurrence.

Preparation

"Run of mine" coal is a mixture of all sizes from 2 or 3 ft across down to dust, along with shale from the floor and roof and dirt bands from within the seam. The coal will be veined with pyrites and other minerals and may even comprise different kinds and qualities of coal. In thick seams these different coals may be separated to some degree at the coal face.

Cleaning the coal is the removal as far as is practicable of the shale, dirt and veined minerals. Apart from this the coal may be prepared for use by separating the different types present, by breaking if necessary to smaller sizes and by screening to standard size ranges. Run of mine coal would probably be sized to ± 3 or

4 in. The large coal would be cleaned by hand-picking and sold for domestic or railway use, or broken to smaller sizes. The undersize would be cleaned by "washing" which would be a gravity separation of coal from shale in equipment similar to those used in ore dressing. Jigs are the commonest type used for coal but riffle tables are also used, and flotation for fine coal. Dry methods—air suspension—are also used occasionally. Coal must be "de-watered" after washing, usually by a draining process.

Breaking is in either pick or roll-type breakers, which should be operated to crack the lumps rather than crush them, to minimize production of fines.

Sizing is by static or vibratory screens or trommels, to standard size ranges known in the trade by particular names:

Large cobbles	− 6 in.	+ 3 in.
Cobble	− 4 in.	+ 2 in.
Trebles	− 3 in.	+ 3 in.
Doubles	− 2 in.	+ 1 in.
Singles	− 1 in.	+ $\frac{1}{2}$ in.
Peas	− $\frac{1}{2}$ in.	+ $\frac{1}{4}$ in.
Grains	− $\frac{1}{4}$ in.	+ $\frac{1}{8}$ in.

The sizes are for round apertures, and the efficiency of screening is specified in terms of the proportion of allowable undersize. Large coal and slack are specified as over or under stated sizes.

The "natural" size distribution from a mine or an area depends on type of coal and mining methods, and is unlikely to be the same as the demand size distribution. Breaking always produces fines so that there may be times when it is not easy to market the whole of production. If fines are in excess, briquetting may be adopted to use them up.

An increasing proportion of coal is pulverized, usually in air-swept ball mills, to sizes about − 100 mesh (mostly much less). This uses up fines and even necessitates breaking down large coal as the demand increases. The main application of pulverized fuel is in power stations and other steam-raising plant. Almost any coal could be used. It need not be clean but it should be dry. Weakly caking steam coal is most suitable. Pulverizing is usually carried out on the site of the furnace or boiler from suitably small material, but pipe-line transport within a works area is quite practicable.

Storage

The storage of coal is not as simple as it might appear. There is a danger of spontaneous combustion due to heat generated by atmospheric oxidation. This danger is greatest in freshly mined coal and is usually serious only in large dumps over 200 tons. The danger is reduced by restricting the height of the stack to 2·5 m by ensuring very adequate ventilation or by stifling it altogether with dust and by water spraying if necessary.

There is also some deterioration during storage over extended periods due to oxidation. There is some loss of calorific value, some reduction in size and an increase in friability. The coking quality may also be impaired. This deterioration is least with large coal. It increases with temperature. It can be mitigated by careful stacking designed to minimize ingress of air.

Uses

The use to which coal is currently being put can be read off Fig. 5. Very little is exported from Britain today and most of that trade is in special qualities. Britain's coal is not, in general, easily won and it is not competitive on the world market. Thick seams in U.S.A., South Africa, Canada and Australia, for example, are cheaper to mine and these countries can put coal on the market at a price which makes long-distance transportation possible.

In 1974 over half the coal used in Britain went to power stations. This was mainly low-grade coal suitable for pulverizing. On an energy basis this probably represents rather less than half the coal energy because of its lower than average calorific value. About 20 per cent of the coal went to the production of coke and other smokeless fuels leaving only about 30 million tons to be used raw. About half of this along with some 5 million tons of coke and smokeless fuels is used in the homes. This leaves only about 12 million tons to be used by industry for a multiplicity of purposes. Much of it will be used for steam raising, for firing kilns and for making producer gas and similar secondary fuels in small quantities. Modern chain grate stokers give well controlled combustion conditions and reduce labour costs associated with handling but these have been replaced time and again by the even simpler to operate oil burners. Pulverized coal may still be used in melting furnaces and billet heating furnaces but again these are easily replaced by oil burning equipment. The trend is away from the direct combustion of raw coal because of the handling costs, the ash disposal costs and the inherent atmospheric pollution problem. Control over combustion rates is not so easy as with gas or oil. Storage of reserves is possible on a big scale but this is costly in terms of space and money. Similar considerations apply to the domestic use of coal where gas, oil, and electricity are increasingly favoured for space heating. Legislation prohibiting the discharge of smoke from chimneys restricts the use of raw coal except in approved appliances. Central heating is now increasingly being installed using oil or gas as fuel and coal fires in modern homes are now being incorporated as architectural features—for the rich!—rather than as the principle sources of warmth and comfort.

The downward trend in the use of coal may now be arrested because of sharp increases in the relative price of oil during the mid-1970s but the development of the North Sea oilfields during the late 1970s and the opening of new mines in Yorkshire and off the East coast where thick easily worked coal seams have been discovered, will undoubtedly affect the pattern of fuel usage during the next twenty years. In any case, the pattern will change as coal gradually takes over from oil early in the 21st century. Whether it will continue to be used for generating electricity then, except in existing plant, is doubtful. Most of it will probably be treated in large chemical processing plants for which the name "coalplex" has already been coined although its exact function remains rather vague. It is envisaged, however, that the needs of the nation for gaseous and liquid fuels will be met by conversion of the coal to whatever chemical species are needed in these plants which will replace today's petro-chemical complexes. The chemistry necessary for this kind of operation is largely understood. The extent to which the idea will become a reality will depend on economic and political factors which cannot be evaluated today

4

Carbonization

CARBONIZATION of coal is its decomposition by heat, out of contact with air, into a solid residue, coke, and liquid and gaseous distillation products.

Types of Process

While the products of carbonization are always coke, gas and liquor, the main objective of the process is usually the production of either hard coke or gas, the liquor being a valuable by-product in each case. The carbonizing industries are then the gas industry and the coke industry, with a small but growing "smokeless fuel" industry taking at present a poor third place.

If gas is the primary objective, coal is used which has a high volatile content and moderate caking power though a coherent coke must be produced. If hard coke is being produced the coal required must have good coking power, found in coals of high rank (88–91 per cent C) and medium volatile matter content (25–30 per cent) (dry ash-free basis).

Coals are usually blended, that is strongly coking coals are mixed with weakly coking ones or with non-coking material like coal breeze or fusain dust. This helps to conserve the rather scarce strongly coking varieties, produces more gas than these varieties would do alone, and is indeed necessary under the fast coking conditions employed today, which would produce very coarsely porous coke if the "best" coking coal was carbonized without dilution.

As coking coal is heated through a narrow temperature range at about 300°C it becomes plastic or even fluid and at this stage most of the decomposition occurs. Gas and tar are evolved and bubbles form in this plastic mass. As the temperature rises further the residue solidifies again and freezes-in the porous macro-structure. Further heating continues to drive off volatile matter and the fine structure develops an increasing degree of order toward that of graphite.

In the process about half the sulphur distils off as H_2S and the remainder is fixed mainly as FeS. Oxygen and hydrogen are largely eliminated and half the nitrogen escapes as NH_3 and C_2N_2.

Carbonizing for gas is carried out in retorts or ovens and may be continuous or static. In the discontinuous process large moulded fireclay or silica retorts are used 20 ft long by about 2 ft across and with a round or ◘-section. These are mounted horizontally in batteries and are heated by producer-gas burned with recuperator preheated air. Flue temperatures may reach about 1300°C but the charge will not all be so high. Each retort holds about a ton of coal which will carbonize in about 12 hr. The coke is pushed through with a ram and water quenched. Volatiles are taken off each retort via an ascension pipe connected at the door on one end and led to a collecting main and thence to condensation and gas-cleaning plant.

Continuous processes are usually conducted in narrow vertical chambers, the walls of which are heated by gas burning in the surrounding flues. Coal is fed at the top and coke withdrawn at the bottom at a rate which determines production rate, and, of course, the form of the heating cycle and hence the coke quality. Steam may be passed through the charge (cf. gas producer, p. 48), to cool the coke and bring about a higher degree of gasification. There are numerous designs of equipment but the principle involved does not change. Gas coke is also made in ovens in a similar way to "hard" coke (see under) but from a different coal blend and using a rather lower temperature.

Metallurgical coke or "hard" coke was traditionally made in

beehive coke-ovens and a little is still made this way. The product was very large-sized and highly unreactive. The simple beehive was slow, inefficient and had no facility for collecting the by-products. Its evolution eventually permitted by-products to be collected and thermal efficiency improved especially when the ovens were arranged in batteries, but the modern oven is much more efficient, productive and easy to work. It gives higher yields and lends itself better to the use of inferior coking coals. The coke produced is usually smaller in size but even this is probably more appropriate to modern needs. The modern coke-oven is a vertical chamber about 4 m high, 13 m long and 35–55 cm wide, with a capacity for about 12–20 tons of coke which is produced in about 10–20 hr.

The oven walls are of silica brick, the high-temperature strength of which ensures a very long life and enables the walls to be built thin for a high rate of heat transference through them.

Ovens are arranged in batteries, each pair being separated by a system of vertical flues in which the fuel gas is burned. Producer gas, blast-furnace gas or coke-oven gas may be used. Air for combustion (and the gas too if lean) is preheated in regenerator chambers beneath the ovens. The temperature in the flues has to be about 1350°C.

There are a number of different oven designs (Otto, Becker, Koppers, Simon–Carves, etc.), which differ mainly in the arrangement of the flues. The main aim is to produce a uniform temperature over all the oven walls in the battery, and the operation of the battery is also directed toward this end. Ovens are charged and discharged in a special order to preserve this uniform distribution of temperature and avoid damage to the battery through a badly distributed load, especially if the coal swells.

The ends of the ovens are sealed with heavy brick-lined doors which are handled by service vehicles which run on each side of the battery. Ovens are charged by a multiple-hopper charging vehicle which runs across the top of the battery and pours the blended coal, hammer-milled to −3 mm, through a series of ports in the oven roof. During coking, gas and tars are taken off via an ascension pipe and passed into the gas-collecting main. Coke, when fully carbonized, is pushed by a ram out of the oven into a quenching car and quenched with water. Proper quenching leaves just enough heat in the coke to dry out the residual water. An alternative "dry quenching" uses circulating gas (N_2 + CO) which transfers the heat to waste-heat boilers; damage to coke is said to be reduced, but maintenance costs are high and the system is not popular.

During carbonization coal swells at first but shrinks as the temperature rises. This can put some pressure on the walls and it is important that an oven should not be pushed until the coke has shrunk clear of the walls. It is also important to blend the charge so that it will eventually do so.

As carbonization proceeds the temperature in the coal rises gradually from the wall inwards, and a narrow zone of plastic coal at about 300°C advances slowly from each wall toward the centre. Behind it there is coke and ahead there is coal. The gas and tar go forward to the middle, then up and out, probably leaving some of the tar behind. A very recent development of practice is to charge preheated coal. This allows faster heating through the critical temperature range, which should accommodate a wider range of coals, and also improve production rate.

Ultimately, the zones meet at the centre which is now only at about 300°C although the outside is at 1300°C. Some time must be allowed for the centre to graphitize, but the outside and inside coke are never quite the same.

The two large sheets of coke in the oven are fissured in a honeycomb pattern, and on discharge break into pieces roughly half as long as the oven is wide and about 15–25 cm across. The outer (cauliflower) ends are hard, silvery grey and finely porous. The inner ends are often softer, almost black, and the pores are more obvious. There may be further variations in appearance, hardness and reactivity with position in the oven, particularly from top to bottom as affected by the mode of charging. The properties of a coke depend mainly on the nature of the coal (blend) but also on operating conditions. The most strongly coking coals (e.g. Durham) give the most unreactive cokes, while the weakly coking coals (e.g. Midlands) produce relatively reactive cokes. Any given coal will yield a more unreactive coke if carbonized slowly, for example, in a wide 55 cm oven, than if coked in, say, a narrow, fast-working continuous vertical retort. High temperatures and long times at temperature also reduce reactivity but induce cracking and may reduce the size of the coke. Coal moisture content, coal size and method of charging also affect the properties, probably by way of their effects on heat transfer through the charge.

Modern practice favours a medium width of oven to effect a compromise between the rather small, but hard and uniform coke from the narrow oven and the higher output of possibly more reactive lumpy coke of the wide oven.

Carbonization, however, remains to some extent an art and a skilful operator, given a reasonable choice of coals and not too hard pressed for output, can usually produce good-quality coke.

Low-temperature carbonization is primarily aimed at the production of smokeless domestic fuels such as "Coalite" or "Furnacite". The temperature used is only 500°C–600°C and low heat transfer rates at so low a temperature are countered by the use of very thin layers, by stirring or by internal heating of the coal charge by producer gas or superheated steam. The fuel produced is reactive, smokeless and suitably sized for domestic use. The yield of gas is low, but a high yield of a tar is obtained which is particularly suitable for adaptation to liquid fuel and is a valuable source of aromatic by-products. This part of the carbonizing industry has become much more important and is still expanding, and much research is being done on finding ways of producing smokeless fuels from a much wider range of coals so that they can be made available in almost any part of the country without incurring high transport costs.

Formed coke. The size and size distribution of coke lumps

determines the burning characteristics of the fuel (see next chapter) and influences the flow of materials through the blast furnace and hence the productivity and efficiency of the iron-making process. It might be expected that if coke could be made to a standard size and shape, better control over its behaviour in the blast furnace could be achieved. "Formed coke" may soon be replacing traditional coke in metallurgical applications not only for the potential improvement it might confer on blast furnace practice but also because it can be made from coals which are mainly or even totally non-coking in character. Formed coke is made in two stages. In the first a low ranking, highly volatile coke of low swelling number (N.C.B. 802, say) is carbonized at a low temperature about 500°C to produce a char. This may be achieved in a fluidized bed or in some kind of retort where rapid heating of the finely crushed charge is possible. Gas and tars must be recovered. The char is blended while still hot with either about 20 per cent of a coking quality of coal or with tar—or with a mixture of these—and formed by pressing or extrusion into briquettes of a suitable shape and size. These briquettes are then heated to carbonize the binder and to consolidate the char to a suitable strength and reactivity. Production of formed coke has reached an advanced pilot plant stage and trials have shown* that its use in blast furnaces is practicable and can be beneficial. Further development of this fuel to a production scale seems certain to take place in the next few years. The exact needs of the users will have to be defined, however, and then means found to meet these needs at a suitable price.

Products and Yields

Table 3 gives typical proportions of the various products of coal carbonization in gas-works and in metallurgical coke-oven practices. The major differences between the two are due to the

* *Journal of the Iron and Steel Institute* **211,** p. 547 (1973).

TABLE 3

Typical Yields of the Various Products of Carbonization of Gas
Coal in Retorts and Coking Coal in Coke Ovens

Product	Gas Retort	Coke Oven
	(weight %)	
Coke	65	75
Gas	22	14
Tar	4	3
Benzole	1	1
Ammonia	1	1
Water	6	5

types of coal treated—high and low volatile coals respectively.
For any given coal blend, however, there may still be considerable
variation in yields due to operating conditions, particularly
temperature and duration of coking, and heating rates. The
operation of the by-products plant will also affect these figures as
the items listed are not sharply defined. The lighter paraffins, for
example, may be divided in varying proportions between gas and
tar.

The primary separation of the by-products is carried out in
several stages: 1975042

1. Cooling of the gas, for condensation of tar and ammoniacal
   liquors.
2. Electrostatic separation of remaining tar.
3. Reheating of gas and washing with sulphuric acid for conversion of ammonia to ammonium sulphate. At this stage
   the ammonia from the liquors (1) are reintroduced to the
   system and also sulphated.
4. Gas is oil washed to remove benzole.
5. Further oil washing for naphthalene.
6. Desulphurization with iron oxide.
7. Drying with calcium chloride or glycerine to reduce the dew
   point.

It will be obvious that this amounts to a gas-cleaning operation and all stages are necessary if the gas is to be sold for general distribution. Gas for internal use may not go through stages 5, 6 and 7. The removal of sulphur would usually appear to be advantageous in metallurgical works but is not always carried out for economic reasons.

While by-product recovery is economically desirable it is also a practical necessity if the gas is to be recovered clean. The overall economic picture is complex and changes from time to time. At one time ammonium sulphate was an important credit, but it can now be produced more cheaply by other means and probably barely pays for its removal. At present benzole production is more profitable. The tar is further treated by distillation to the technical fractions naphtha, light oils, carbolic oil, creosote oil, anthracene oil and pitch. The exact fractionation applied depends on demand. Further splitting up of these fractions for rarer or more valuable aromatic compounds is not likely to be undertaken at the coke- or gas-works.

The properties of the gas produced are discussed in Chapter 6 and those of the tarry liquors in Chapter 7.

5

Coke

Classification

There is a clear distinction between gas coke and metallurgical or "hard" coke. The former is smaller, weaker and more reactive. Its chemical composition is not closely specified and its pore structure is likely to be rather open. It is well suited for domestic appliances for which it should be closely sized and as free as possible from inclusions of shale which cause it to spark. It can also be used in small central heating plant and in gas producers.

Hard coke is made primarily for the metallurgical industries and particularly for foundries, for use in cupolas and crucible melting furnaces and for blast furnaces. The traditional beehive coke was very large and unreactive and afforded a means of attaining very high temperatures. Modern coke is smaller and perhaps rather less unreactive, but meets present-day requirements well.

On leaving the ovens, coke is quenched with water, dried and then screened. The smaller sizes (below about 5 cm) are sold for domestic and general purposes, and only larger sizes used in blast furnace or foundry. Obviously the yield of large-sized coke should be high at this stage and further breakdown prior to use should be as small as possible. "Domestic" coke, down to about 5 mm, is less valuable than "blast furnace" coke. Breeze, below 5 mm, is not easy to get rid of in large quantities, but a certain proportion is necessary as part of the change to the sintering machines in modern integrated iron and steel plant.

Properties and Tests

The relevant properties of coke will be discussed under the following headings:

1. *Chemical*
   - (a) Analytical — Proximate—H_2O, S, ash, volatile matter and fixed carbon.
     Ultimate—C, H, N, O and ash analysis.
   - (b) Reactivity — to CO_2 and O_2.
   - (c) Calorific value

2. *Physical*
   - (a) Strength — Shatter tests, drum tests, size analysis.
   - (b) Analytical — Real density, apparent density, bulk density, porosity.

All of these properties may be examined in either gas coke or hard coke, but the discussion which follows concerns mainly the properties of hard coke as required in iron-making as it is here that the highest quality is required and testing is carried out most rigorously.

1. *Chemical Tests*

The effects of chemical analysis on the metallurgical process are largely calculable. *Moisture* should be down to below 3 per cent. It must not be paid for and charged to the furnace as carbon. *Sulphur* in coke becomes part of the metallurgical load, entering metal and slag. In iron-making 90 per cent of the sulphur comes from the coke, and much of it must be removed as CaS necessitating an addition of limestone and a corresponding amount of coke to heat and melt it. The sulphur content of coke is usually about 1 per cent. It should be as low as possible but users often prefer a steady value even if it is always higher than the minimum occasionally attainable. Similarly *ash*, which is present at about 10 per cent, should be low, but steady. Ash is largely silica and alumina and must be fluxed in the blast furnace with lime which in turn must be heated and melted. *Volatile matter* is not available as fuel and should be low (1 per cent), especially as this would indicate thorough coking. *Fixed carbon* is used for estimating the calorific value of the fuel and should be about 85 per cent or more if the ash is low.

The *ultimate analysis* is seldom carried out as a routine, because it is tedious and gives information which probably doesn't change much from day to day and tells little more than the proximate analysis. Typical values on a dry, ash-free basis would be:

	%
Carbon	95
Hydrogen	1
Nitrogen	1
Sulphur	1
Oxygen	2

It should be understood that most of the sulphur will be as FeS—the Fe appearing in the ash analysis.

The determinations of *reactivity* (to CO_2) and of *combustibility* (to O_2) are usually research jobs. The only standard test is the "Critical Air Blast" (C.A.B.) test which is a test of "ignitability". This is carried out by determining the minimum air flow rate through a bed of fuel, in a standard piece of equipment, which will maintain its combustion once started. Values obtained in this test range from 0·01 ft³/min for highly reactive gas cokes to 0·07 ft³/min for very unreactive metallurgical coke. Less official tests involve comparing the rate of weight loss of standard samples of fuels under standard conditions, in atmospheres of air, oxygen, or CO_2. The results obtained depend very much on the conditions under which these tests are carried out—such as the temperature, pressure and gas flow rate—and the reaction rate alters during the test.

Any distinction between reactivity and combustibility is probably false since those factors which affect one will always affect the other in a similar way. There is, however, a hope among users that a coke which is extremely unreactive to CO_2, but extremely combustible in air, will be developed. This seems most unlikely ever to happen.

Reactivity depends on the carbonization conditions—mainly temperature and time—on the pore structure of the coke to some extent and probably on the nature and amount of minor impurities in the material which can exercise a catalytic effect. The reactivity of very pure char is known to be very low, but it is increased by even small additions or iron or sodium oxide (among other things) probably according to mechanisms such as:

$$2Fe + O_2 = 2FeO$$
$$2FeO + 2C = 2Fe + 2CO$$
and
$$Na_2O + CO_2 = Na_2CO_3$$
$$Na_2CO_3 + C = Na_2O + 2CO$$

Coke is sometimes sprayed with sodium carbonate solution to enhance its reactivity for domestic use.

The *calorific value* of coke is rather lower than that of anthracite or even of bituminous coals. This is partly because of the lower hydrogen content and also because of the higher ash content (see Table 1).

2. Physical Tests

Coke strength is usually measured by subjecting a sample of large coke to standardized abuse and sizing the products on standard screens. Occasional reports have been made of tests designed to measure the strength or hardness of the coke substance rather than that of the coke lumps—hardness and abrasion tests. These might be of significance in research but cannot describe the initial tendency of coke to break down to small sizes as this is very dependent on the system of cracks which develops in the oven or during cooling.

The Shatter Test was for long the standard test in Britain. Fifty pounds of $+2$ in. coke were dropped four times from 6 ft on to an iron plate and sized on 2 in., $1\frac{1}{2}$ in., 1 in., and $\frac{1}{2}$ in. square-mesh screens. This test was originally developed to show the probable size distribution on arrival in the blast furnace in a typical works, presuming the sample to be representative of the coke on entering the works at coke-oven screens or railway sidings.

Good metallurgical coke (e.g. South Wales and Durham) would give a size analysis after shattering something like 80, 90, 96, and 98 per cent, while a poor coke (Yorkshire) would give 65, 80, 90 and 95 per cent. If the $-\frac{1}{2}$ in. fraction in the Yorkshire coke fell to 3 per cent, this coke would be considered remarkably good, but if in a Durham coke it rose to 3 per cent it would be considered poor. In fact, it is the day-to-day variations that are important *within* a coke plant rather than the comparison with incomparable plants, and these day-to-day variations can be considerable, depending on coal blend, oven temperatures, carbonizing time and other factors. Users prefer regular quality.

Attention tends to be focused on the $+1\frac{1}{2}$ in. and $-\frac{1}{2}$ in. figures and the latter is often said to be the most important for blast-furnace operation as such small coke is most unwelcome in the furnace. Small sizes are screened out before charging but if the coke is weak more may form within the furnace.

The alternative *Micum Trommel Test* was imported from Europe about 1945. In this 50 kg of large coke $+50$ mm was rotated in a drum, with internal lifting shelves, for 100 revolutions in 4 min and then sized on round-holed screens at 100, 80, 40, 20 and 10 mm diameter, the oversize in each of these being quoted. Attention was paid mainly to the $+40$ mm and -10 mm fractions. This test often fails to agree with the Shatter test in its assessment of coke strength. Abrasion is taken more into account by it but the main difference is that it tests the coke after a much greater degree of abuse than does the Shatter test—in fact to the equivalent of about 40 drops from 6 ft.

In the *Cochrane Abrasion Test* 25 lb of coke is rotated in a small drum to 1000 revolutions. The percentage $+\frac{1}{8}$ in. is quoted as an abrasion index and should be about 80. This test involves even more abuse than the other tests.

If coke is subjected to an increasing number of equal "blows" its statistically determined average size $\bar{x}$ is found to decrease linearly as the logarithm of the number of blows increases (Fig. 2).

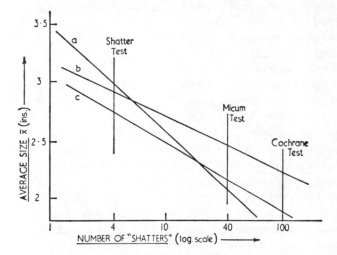

FIG. 2. The effects of the Shatter test, the Micum test and the Cochrane test on the statistically average size $\bar{x}$ of (a) a large weak coke, (b) a large strong coke, and (c) a smaller strong coke. It can be demonstrated that the 100 turns of the drum in the Micum test does damage equivalent to 40 Shatters and that 1000 turns in the Cochrane drum is equivalent to 100 Shatters. Note how (a) and (b) are assessed differently by the Shatter and Micum tests.

This means that any strength test like those described is very sensitive to previous history and a sample which has already been roughly handled is likely to appear strong in the test. This is because much of the breakdown is by the propagation of pre-formed cracks. Once these are used up new cracks are not readily formed.

The initial breakdown, as in the Shatter testing of "fresh" coke, is apparently very severe, but that of well-handled coke is less so.

The difference in the breakdown in the Micum and Cochrane tests between fresh and partly broken coke is not so great as in the Shatter test. If the initial size is large but the slope in Fig. 2 steep, the Shatter test will indicate good coke, but if the initial size is small and the slope is shallow (a much better coke) the Shatter test might show the coke to be relatively poor. In these cases Micum tests might give quite the opposite impression if the lines crossed between the Shatter and Micum ordinates, as do (a) and (b) in Fig. 2.

The tests are obviously not satisfactory except as control tests on fairly regular production. An alternative scheme of testing has been suggested whereby the actual size at the blast furnace is determined and possibly a further breakdown test applied here to simulate damage within the furnace.

The determination of an average size parameter as used in Fig. 2 is not easy because coke sizes are distributed in a bimodal distribution. The parameter $\bar{x}$ already mentioned referred to the modal size of the large material, and ignored the 5 per cent or so of "fines" which become important only after much damage has been inflicted.

Real density is determined by water displacement in a density bottle and indicates the thoroughness of coking—the degree of graphitization. The value should approach, or slightly exceed, 2·0. This quantity is not readily determined with accuracy. The sample must be finely pulverized to ensure access to the whole pore volume. The coke must be thoroughly degassed by evacuation before the water is admitted to fill it up. Other displacement fluids give slightly different results as they are adsorbed to different degrees on the extensive inner surface of the coke. Helium displacement, a special technique, gives the highest and probably truest value, as helium is not adsorbed on the coke surface but can readily effuse into the innermost pore space.

Apparent density is the mass per unit volume of coke including pore space. This must be determined on lumps and good reproducibility is impossible to attain as lumps can be very different one from another. The usual method involves soaking a known large mass of coke in paraffin oil and then determining its volume by water displacement, assuming that the oil stays in the coke and properly defines its limits. This method leaves much to be desired in absolute accuracy. The best method is to cut prisms of coke, weigh each, seal off the pores with wax and determine volume by displacement. Each determination is then accurate but adequate replication is needed to yield an accurate mean. The apparent density of coke is about $1 \cdot 0$ g/cm^3 (1000 kg/m^3).

Porosity is calculated from real and apparent densities and its accuracy is subject to the errors in each of these. Its value lies in the range 45–55 per cent and is often close to 50 per cent. It was once considered that porosity might affect reactivity but the total range is so small that the effect would have to be through pore-surface area, which depends on pore-size distribution. These areas do vary between cokes but the largest specific surface areas encountered are probably not much greater than twice those of the smallest. Porosity is now believed to have only a minor effect on reactivity.

Bulk density is the mass per unit volume including voidage between lumps, and is usually determined by weighing a level wagon-load. It is of interest as a factor for converting volumes to weights when charging is on a volume basis. Corrected to a standard apparent density it is also a sensitive index of strength— or rather of degree of breakdown—and one authority has claimed that coke bulk density correlates better with blast-furnace performance than any other property.

The quality of coke is very much tied to the nature of the coal from which it was made and the results of these tests are barely adequate to describe the differences between cokes made from South Wales, Durham, Yorkshire and Scottish coals. Under test or under the microscope they often seem rather similar while the practised eye or ear can easily tell that they are quite different.

6

Gaseous Fuels

GASEOUS fuels are popular both industrially and domestically for their cleanliness and flexibility. In Britain most of the gas used is distributed by the British Gas Corporation through Area Gas Boards. There is a considerable seasonal fluctuation in demand which it is difficult to match to a more or less steady supply rate. Massive underground storage, ideally in spent oil wells is the most satisfactory way of ironing out these difficulties so that peak demand rates can be satisfied out of savings rather than by trying to step up production rates. Very large users, especially the steel and petrochemical industries make some or all of their own gas, producing to their own requirements a cheaper if sometimes cruder fuel.

Gas has the advantages over solid fuel that there is no ash to dispose of (unless the gas is made by the user) and there should be no smoke produced. It can be tapped in large or small quantities and if bought from the Gas Board it can be expected to be of regular quality and constant calorific value. It can be burned by a variety of techniques to short hot flames or long cool lazy flames to meet a wide range of requirements.

Classification

Gaseous fuels are available from a variety of sources:
1. Natural—from oil and coal measures.
2. From carbonization plants.

3. From "gas producers", from coal or coke.
4. By cracking of oil.
5. From blast furnaces.
6. By bacterial decomposition of sewage (for use in sewage plant).

Of these, natural gas has now become the most important gaseous fuel in many parts of the world. Steel works produce large but not necessarily increasingly large quantities of coke oven gas most of which is consumed within the industry. Other manufactured gases whether from coal or oil are of diminished importance today but natural gas will not last long at the current rate of use and the gas industries will be obliged, if they are to survive, to revert to the manufacturing of gas within three or four decades.

Production

Natural gas has been exploited most fully in the U.S.A. where it has been piped from the oil fields in Texas and Louisiana for distribution in most parts of the country. Rapid development of this industry dates only from about 1945. By 1967 natural gas was supplying about one-third of the energy needs of the country but the production rate is now falling off and the demand created in times of plenty can now be satisfied only by importing similar gas from Canada and Mexico with the possibility that sea-borne imports of liquefied gas may be made in the future. In Europe natural gas has been used in the Po Valley since the turn of the century but only since 1950 has this field been exploited vigorously. By 1960 it was supplying 10 per cent of Italy's energy needs. In France a field at Lacq in the Pyrenees was opened in 1956 and in 1960 the Dutch found the biggest gas field in Europe in the northern Netherlands at Slochtern. This provides the Dutch with a third of their own energy needs and they export a similar amount to Germany whose needs are far greater than can be met from her own smaller and more scattered resources.

By 1964 Britain was importing liquefied natural gas from Algeria, and still does so, having developed refrigerated tankers for the purpose. In 1965 gas was found under the southern end of the North Sea in the sector in which Britain had acquired exploitation rights by an international treaty. This has been found in geological formations similar to those in which the Dutch gas was found. Ten years later this gas provides 15 per cent of Britain's energy needs and 90 per cent of her distributed gas is natural gas. Meantime countries not so favoured are buying it as "lng" from oil producing countries, as is Japan from Libya and Indonesia, or they are being supplied overland by pipeline as is Austria from Russia.

Natural gas is invariably formed along with either coal or oil and is always found in association with these—as firedamp in coal mines and in solution in the petroleum or overlying the oil in the wells. Gas may become separated from the oil during geological folding of the associated rock formations so that detached bodies of gas may be found as is the case in most of the European fields.

The methane of coal measures can be tapped off prior to mining. This is most commonly done when the amount of methane present makes the coal dangerous to mine. The gas is pumped off through pipes driven into the coal ahead of the face being worked. It can be used economically locally or fed into natural gas grids but the amount so recovered would never make a really significant contribution to the economy of a nation.

Dissolved gases in crude oil are always drawn off under reduced pressure to render the oil safe to handle. This may be destroyed by burning (flared off) or it may be liquefied and sold or it may be distributed as natural gas by pipeline. Some oil wells have large bodies of gas present in association with the oil. This is called "associated" gas or sometimes "wet" gas and it is contaminated with lower liquid paraffins. Gas from detached pockets is not so contaminated and is referred to as "dry" gas. Texan gas is mainly associated while North Sea gas is mainly dry.

Natural gas is not usually pure methane but would normally contain about 90 parts of methane to 5 or 6 parts of heavier hydrocarbons ethane, propane and butane along with small amounts of CO_2, N_2 and H_2S. The range of composition is very wide however. Nitrogen in some Dutch gas rises to about 50 per cent. Carbon dioxide can rise to 20 per cent and hydrogen sulphide to 17 per cent. Such gas is described as being "sour" and the Lacq gas can have up to 30 per cent of these two gases so that only 70 per cent of the gas extracted can be sold. Clearly it is necessary to adjust the composition of crude gas to bring it within a suitable specification. In all cases H_2S must be removed (and a more acceptable "malodorant" added), CO_2 must be reduced to no more than about 2 per cent and nitrogen also reduced to a standard level. This is necessary in order to standardize combustion characteristics and calorific value. When natural gases from different sources are being mixed it is essential that these properties be moderated to avoid difficulties for the users. This is particularly important in Europe where gases come from so many different sources.

The transmission of natural gas by pipeline over very great distances is now well established. The cost is not low and it is desirable that the largest diameter of pipe be used that can be kept supplied in order to minimize pumping costs. The pipeline pressures are usually in the range 2000–4000 kN/m^2 (20/40 bar). Pumping stations are required every 200–300 km. Transportation by tanker is usually even more costly as it involves not only special ships but refrigeration and re-evaporation plant and storage capacity to act as a buffer between intermittent arrivals and constant demand.

Producer gas is manufactured by total gasification of coal or coke in a gas "producer" or "generator". This is a refractory-lined cylindrical reaction vessel fed with solid fuel at the top against a stream of ascending gas derived from air or steam or both, injected or drawn in through the bottom. The reactions take place at about 1000°C or over and the gas produced is drawn off at

700–800°C. The important reactions which occur are:

(1) The "producer gas reaction".

$$2C + O_2 = 2CO + 53,200 \text{ cal } (222 \cdot 7 \text{ kJ}).$$

(2) The "water gas reaction".

$$C + H_2O = CO + H_2 - 32,300 \text{ cal } (135 \cdot 2 \text{ kJ}).$$

Actually (1) occurs in two stages:

(1a) $\qquad C + O_2 = CO_2 + 94,200 \text{ cal } (394 \cdot 3 \text{ kJ}),$

(1b) $\qquad CO_2 + C = 2CO - 40,800 \text{ cal } (170 \cdot 8 \text{ kJ})$

and the bed must be deep enough that the second is almost completely accomplished. If the fuel bed is not hot enough, i.e. below about 800°C, (1b) will go too slowly and a part of the carbon will not be available as a gaseous fuel. The producer gas reaction yields a theoretical product 67 per cent N_2 and 33 per cent CO accompanied in practice by a little CO_2 and hydrogen, methane and tarry matter in quantities depending on the nature of the coal or coke gasified.

Reaction (1) is strongly exothermic so that the producer runs very hot. The gases come off at up to 750°C and are often used immediately as in open-hearth steel furnaces to utilize the sensible heat. The fuel bed also becomes very hot and clinkering of ash is likely to cause inconvenience in its withdrawal at the bottom of the producer.

The excess of heat over that needed to maintain a reasonable temperature may be absorbed by water jacketing and running the unit as a waste heat boiler. More usually steam is injected with the air so that the endothermic water gas reaction (2) can absorb the excess heat. The calorific value of the gaseous product is raised as nitrogen is replaced by CO and H_2 in equal proportions.

The optimum proportion of steam to air is primarily decided by the thermal balance but side effects arise if the proportion of steam injected is too high. Particularly, the temperature of the bed falls and the CO_2 content of the gas is increased by the reaction:

$$CO + H_2O = CO_2 + H_2$$

as the partial pressure of H_2O rises. The H_2O content of the gas produced becomes unduly high and its calorific value falls. The amount of steam used is usually about 0·4 kg per kg of fuel unless severe clinkering makes cooler operation essential.

Gas producers may be fed with almost any carbonaceous fuel but the important choice is between coke and anthracite on the one hand, which give permeable fuel beds and clean gas, and weakly caking bituminous coal on the other, which does not swell and choke the fuel bed, but yields a tarry gas burning with a good luminous flame. If impermeable beds are formed air should be supplied under pressure and the hearth rabbled to maintain gas flow. The producer coal should be well sized for good permeability coupled with large reacting surface area.

Producer gas with its high nitrogen content (see Table 4) has a very low calorific value and, when burned with air, provides low flame temperatures unless considerable preheating of both gas and air is employed. Producer gas made with oxygen enriched air and steam would have a much higher calorific value. The generator would run at a higher temperature but some of the heat could be put toward steam-raising. Producers using oxygen and steam are operated so that the coal ash is melted or slagged for easy removal from the generator. The gas obtained from these producers is usually rather high in CO_2 and its calorific value is only about 2000 kcal/m^3 (8000 kJ/m^3)·

Water gas is sometimes required undiluted with nitrogen, particular applications being in reheating furnaces for forging and forge welding of steel. Water gas generators operate by a two-stage cycle. Air is blown through the fuel bed to heat it up beyond 1000°C, the products of this stage being passed to the chimney. The second stage is to blow steam through the incandescent coke until the temperature falls to about 900°C, the gas being fed to a gas holder which is necessary to accommodate an intermittent supply to a constant demand. When the temperature falls to 900°C the steam is cut off and air again blown.

The product is known as "blue" water gas. Its uses are rather restricted but it can be more easily used to higher temperatures than producer gas and it is also used as a source of hydrogen for the chemical industries.

Carburetted water gas is a more important fuel; it is enriched with cracked oil vapour. In this case, the producer gas made during the air blow is used to heat up a carburettor and cracking chamber. This latter is a hot box of chequer brickwork, held at about 800°C. During steaming, oil is sprayed into the water gas stream as it enters the cracking chamber and about three-quarters of it is decomposed to variable mixtures of methane, ethane, unsaturated hydro-carbons and hydrogen. The remainder forms tar which would usually be recovered, and char which helps to heat the chequers in the heating stage following. The gas oil used is a light fraction boiling at about 200°C. The ratio of coke to oil is about 3 : 1 by weight. The gas produced has an enhanced calorific value and can be used as a component of town's gas. This type of generator is flexible in operation, and by replacing coke with oil reduces the demand for coking coal for gas production.

Oil gas is made by a similar process but from heavy oil which may require to be decomposed catalytically in the carburettor. It may or may not be mixed with water vapour. Its calorific value is very high. It has been used as a town's gas but has been displaced in the U.S.A. by natural gas, and for other purposes by bottled gas.

Blast-furnace gas is a low-grade producer gas available in vast quantities at iron-works. Thoroughly cleaned of dust it is used extensively in integrated steel plant for Cowper stoves, steam-raising for power, reheating furnaces and even in gas engines for driving blowers or electricity generators. Modern blast-furnace practice tends to reduce still further the calorific value of this fuel but the total potential heat is still large and the gas will continue to be utilized even if it has to be enriched slightly with coke-oven gas. The most recent developments in blast-furnace practice—fuel injection, oxygen enrichment and steam injection—may result in an increase in the calorific value of the top gases mainly due to a fall in nitrogen and a rise in hydrogen content.

Retort gas is the gaseous product of carbonization of weakly coking gas coals. It is rich in hydrogen and methane and has a high calorific value and as the traditional basis of town's gas was formerly widely used both in homes and by industry. "Steamed" retort gas has a higher carbon monoxide content and a slightly lower calorific value. The toxicity of the carbon monoxide is the greatest disadvantage of this gas especially in view of its very wide distribution.

Coke-oven gas is similar in composition to retort gas and is used in the steel industry for many purposes. It cannot be preheated because the methane would crack in the regenerator chambers and deposit soot. It burns with a clear, non-luminous flame with poor radiation properties and must be used with some oil to produce luminosity. Open-hearth furnaces were often fired with a mixture of coke-oven gas and oil or coal-tar fuel or pitch.

Refinery gases are the light hydro-carbons fractionated from petroleum oil. These are the C_2, C_3 and C_4 hydro-carbons, but some C_3H_8 (propane) and C_4H_{10} (butane) would normally be separated for sale as "bottled" gas. The residue could be mixed in with town's gas either unchanged if the addition was to be small, or as *reformed gas* if a large addition was to be made. Reforming is a partial combustion with the aid of alumina as a catalyst at about 850°C:

$$2CH_4 + O_2 = 2CO + 4H_2$$

Butane and *propane* from petroleum are liquefied and distributed in cylinders for use in homes, caravans and to an increasing extent industrially, e.g. for welding. Their calorific values are very high and very hot flames can be produced using them.

Non-explosive butane-air mixtures can be safely piped for use in compact communities if suitable burners are used (see page 57).

Town's gas supply is traditionally based on gas retorts but usually a part of the coke produced is gasified to water gas or carburetted water gas for mixing with the retort gas. Any of the other gases mentioned might be added in also, provided the declared calorific value is maintained and the gas supplied can be used safely in the standard burners. Town's gas is distributed in accordance with a number of legal safeguards and regulations. These cover toxicity (H_2S and HCN but not CO) and calorific value particularly. It is available for both domestic and industrial use but many industrial users find it too dear and prefer to make their own gas, probably less pure but equally satisfactory. Some fortunate industrial areas have a gas grid through which all large gas users and makers are linked to their mutual benefit. Distribution is no longer under control of the "town", of course, but is the responsibility of the Area Gas Board. Most distributed gas is now natural gas and the term town's gas is virtually obsolete.

Complete gasification of coal in one stage to a high calorific value gas suitable for use as a town's gas has long been the aim of several processes. One process which has had some success even with low-grade coals is the *Lurgi* process which may have a place in future developments. A full-scale "experimental" plant is now operating in Fife and supplying the Scottish gas grid. Current opinion is that this experiment will not be followed up because gas derived from oil and later natural gas, have become cheaper alternatives.

Fine coal is gasified in a producer operating at 1000°C and 20 atmospheres pressure by reaction with an oxygen steam mixture. The product is like steamed retort gas but with a higher CO_2 (30 per cent) content which must be scrubbed out. It contains about 15–20 per cent of methane and other hydro-carbons which are probably formed by hydrogenation of the tars under pressure. At higher temperatures the CO_2 might be reduced to smaller proportions but clinker formation limits the temperature that can be used. The great benefit of such a system is that low-grade and non-caking coals can be transformed into high-grade gaseous fuel.

Among minor sources of gaseous fuel the most interesting is underground gasification which would reduce the labour of mining if developed to complete success. Air, or air and steam, are pumped into thin unworkable seams and made to react as in producers. The gas drawn off from an adjacent part of the seam is a lean producer gas. A higher calorific value could be obtained by using oxygen-enriched air with steam. Various schemes to develop these ideas have shown some promise but none has yet led to large-scale operation.

Properties and Uses

Typical analyses of gaseous fuels are presented in Table 4 with their calorific values. They fall into three distinct groups. The straight hydro-carbons derived from petroleum have very high calorific values ranging from 9000 kcal/m^3 for gas which is essentially methane, up to nearly four times that figure where the major constituent is butane. The difference is, of course, mainly due to density, the C_4 hydro-carbon being about four times as dense as methane when in the gaseous state. Expressed as calories per gramme these figures all become similar to those for liquid fuels, gasoline and kerosene in Table 5.

Those manufactured gases which are prepared so as to be low in nitrogen have calorific values about 3000–4000 kcal/m^3 and are the ones which are normally referred to as "rich" to distinguish them from the lean fuels with 50–60 per cent nitrogen and calorific values around 1000 kcal/m^3.

Two important, related properties of gaseous fuels (and of liquids burned after vaporization) are the *velocity of flame propagation* and the *inflammability range* which is bounded by the upper and lower inflammability limits of the gas-air mixtures expressed as the volume per cent of the fuel gas. Flame velocity depends upon the activation energy of the combustion reaction and on the capacity of the heat evolved at any stage to raise the

TABLE 4

Typical Analyses of Fuel Gases in Per Cent by Volume and their Approximate Calorific Values

	Analysis								Net Calorific Values		
	H_2	CO	"CH_4"*	"$CnHm$"*	CO_2	O_2	N_2	Others	kcal/m³	B.t.u./ft³	kJ/m³
Natural gas	—	—	90	C_2 3 C_3 1·5 C_4 0·5	1	—	3	—	9000	1000	37,670
Refinery gas	8	—	10	C_2 10 C_3 45 C_4 20	—	—	—	H_2S:7	18,000	2000	75,350
Domestic bottled gas	—	—	—	C_2 3 C_3 22 C_4 75	—	—	—	—	30,000	3400	125,600
Coke oven gas	54	7·5	28	2·5	2·5	0·5	6	—	4000	450	16,740
Retort gas (coal)	52	9	28	3	2·5	0·5	5	—	4250	475	17,790
Retort gas (steamed)	50	18	20	2	4	0·5	5·5	—	3800	425	16,000
Blue water gas	49	41	1	—	5	0·5	5	—	2500	275	10,400
Carburetted water gas	37	30	14	7	6	—	5	—	4000	450	16,740
Oil gas (steamed)	48	19	14	6	8	0·5	4·5	—	4450	500	18,830
Producer gas (coal)	12	29	2·5	0·4	4	—	52	—	1350	150	5,600
Producer gas (coke)	11	29	0·5	—	5	—	54·5	—	1100	125	4,600
Producer gas (air only)	1	33	—	—	1	—	65	—	900	100	3,770
Blast furnace gas	1	27	—	—	11	—	61	—	800	90	3,350
Lurgi gas (scrubbed)	50	19	22	1	7	—	1	—	4000	450	16,740

* Methane may include a little of other paraffins. The composition of the unsaturated hydro-carbons varies from one fuel to another.

temperature of the next increment of the mixture to the ignition temperature. The flame velocity rises from zero at the lower inflammability limit through a maximum when air is below the stoichiometric proportion and back to zero at the upper limit. Below the lower limit there is not enough fuel to provide the heat to raise the temperature of the excess of air present to the necessary level. Above the upper limit the heat evolved is limited by oxygen supply and cannot heat up the excess of fuel gas present to the ignition temperature. (See Fig. 3.)

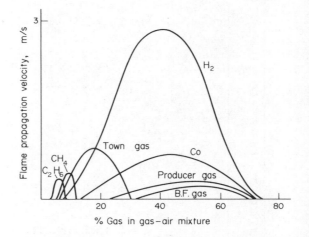

FIG. 3. Comparison of flame propagation velocities and inflammability limits for a selection of gases mixed with air. Note that the actual velocity measured is dependent upon the characteristics of the measuring equipment.

Hydro-carbons, and particularly high hydro-carbons, require very large volumes of air to burn one volume of gas. Consequently only small excesses of either gas or air can be tolerated and the inflammability ranges are low in these gases—6–12 per cent for methane, and 4–9 per cent for ethane. Hydrogen, on the other hand, needs only $2\frac{1}{2}$ times its own volume of air for complete

combustion. Its combustion kinetics are very simple compared
with those of the hydro-carbons. As a result flame velocities are
very high, under favourable conditions, and large excesses of
either hydrogen or air can be carried, the inflammability range
being about 8–70 per cent. Carbon monoxide requires the same
volume of air as hydrogen. Flame velocities are much slower
than with hydrogen but the inflammability range is only a little
narrower at 17–70 per cent. The additional nitrogen in producer
gas increases the mass of inerts to be heated up so reducing the
flame velocity and the inflammability range, the latter to about
25–60 per cent. Obviously the inflammability range will be
extended in all cases if the air is enriched in its oxygen content.
The actual figures obtained for flame velocity or for the limits
depend on the conditions under which measurements are made—
including temperature, pressure and the configuration of the test
equipment used.

Either inflammability limit can be estimated for a mixture of
fuel gases with air by applying Le Chatelier's Rule that:

$$\text{The limit for the mixture} = \frac{a + b + c + \ldots}{a/A + b/B + c/C + \ldots}$$

where a, b, c are the volume percentages of each gas in the gas-air
mixture, and A, B, C are the limits for the respective gases.

These properties most obviously affect explosion risk, flame
behaviour, ignition characteristics and burner design. Gaseous
fuels cannot necessarily be interchanged or intermixed without
some adjustments being made to equipment or to practice. The
important consequence of this in recent times has been the need
to "convert" domestic and industrial gas fired appliances for use
with natural gas. This has been carried out area by area and
street by street across the whole of Britain at considerable cost.
The modified burners have had to be designed to use gas at about
half the volume rate (to allow for the higher calorific value) but
to be supplied with air at twice the rate. The gas linear flow rate
had to be further reduced, however, to allow for the lower flame

propagation velocity and where possible a greater space created for the development of the flame. Even between one quality of natural gas and another, adjustments sometimes have to be made. Sometimes it is sufficient to change the delivery pressure by a few mbar but even this is not an operation which can be undertaken lightly.

Producer gas and coke oven gas also need quite different arrangements, not only because of the above considerations but because producer gas must be preheated to provide a reasonably high flame temperature whereas coke oven gas must be fired cold because methane in it would crack at preheating temperatures. In fact a certain amount of mixing is possible, however, and gases have in the past frequently been blended for public distribution, the burning characteristics being controlled by maintaining a suitably high proportion of hydrogen in the mixture.

Gas continues to compete successfully with oil and electricity in many industries. It is clean and can be almost completely de-sulphurized if necessary. It need not require on-site storage facilities, warming equipment or injection steam like oil. Gases can be burned with extreme intensity and can give very high heating rates particularly if convective transfer of heat to the charge can be tolerated. The use of methane or propane with oxygen in "fuel lances" is being developed. These new techniques produce something like electric arc conditions from the combustion of a gas.

The products of combustion of gases are a vehicle for the transfer of waste heat to recuperators, and here is a positive advantage over electricity. Gases are also used, only partly burned, as reducing atmospheres in metallurgical furnaces to prevent oxidation or decarburization of stock and even as reducing agents in extraction processes.

The uses of gas for heating purposes are without number, and range from domestic heating up to steelmaking. It is also used as the source of power in gas engines which may produce electricity or drive locomotives. It can take part in metallurgical reactions or may be used as the starting point—as crude hydrogen, carbon monoxide or methane—in many heavy chemical processes. The oil industry has certainly invaded many of the preserves of the gas industry in recent years but it would appear that as the economy expands both kinds of fuel will continue to be required together for a long time.

7

Liquid Fuels

Classification

Liquid fuels are mainly oils, tars and pitches and are derived from the following sources:

1. Petroleum
2. Oil Shales
3. Coal, by carbonization
4. Coal, by hydrogenation

At present petroleum provides the majority of our liquid fuel requirements. The reserves of oil shales are immense, however, and these will probably increase in importance in the future. The liquid and tarry products of carbonization are used by industry near coalfields and carbonizing centres and, as a result, the steel industry employs these in fairly large quantities. Hydrogenation of coal has been exploited mainly by the Germans using lignites. It competes economically with oil only with great difficulty but the process has had strategic significance.

Production

Petroleum is available in many parts of the world and new sources of supply still continue to be found. The crude oils from various parts of the world have a lot in common and their ultimate analyses do not differ greatly from one oilfield to another, with carbon at about 84 per cent $\pm$ 3, hydrogen 13 per cent $\pm$ 1, sulphur up to about 3 per cent, and perhaps 0·5 per cent each of nitrogen and oxygen. All these crude oils consist of mixtures of hydro-carbons—paraffins, olefines, naphthalenes and aromatics ranging from the simplest gaseous members of the homologous series to complex waxes and bitumens—all interdissolved.

The crude petroleums do differ, however, in that some are predominantly paraffins and distil to leave a residue of paraffin wax, while others are mainly aromatic and yield an asphaltic or bituminous residue on distillation. Others have a lot of naph-thalenes present, and every classification is based on the propor-tions of these three components. These differences are important to the refiner who has to sort out the components by distillation into useful groups and if necessary modify these chemically, so that the yields of the various fractions roughly match the demands of industry for them. It is not, however, suggested that all petroleum is divided neatly into a small number of sharply defined fractions. There is obviously much overlapping and room in which to manoeuvre to satisfy the market. The market is not, however, as regular as the composition of petroleum and other operations, apart from fractional distillation, are applied either to break down heavy hydro-carbons into lighter ones (cracking) or to build the lighter ones into others with larger molecules (polymerization). These operations can also modify the propor-tions of different types of molecules, long chain, short chain, branched and aromatic ring types for example, so modifying the properties of the fuel in any boiling range bracket. A high octane number (anti-knock index) in petrol is obtained if the proportion of branched molecules is high, for example.

Refining is the fractionation of the petroleum into its com-ponents and the first stage is a simple fractional distillation as follows:

1. Natural gas—Boiling range below 30°C. This is usually removed under reduced pressure at the oilfield, initially to permit safe handling of the crude oil.

2. Gasoline—Boiling range 30–200°C. This is "petrol" and may be further divided into aviation spirit (30–150°C), motor spirit

(40–180°C), and vaporizing oil (110–200°C) for tractors, etc.

3. Naphtha—Boiling range 120–200°C. There is usually a surplus at the top end of the gasoline range which is used for further processing. This is the fraction which is reformed to high octane gasoline and supplied to petrochemical plants as feedstock. It was also used for making into town's gas during the 1960s.

4. Solvent spirit—Boiling range 120–250°C. This is white spirit or turpentine substitute and is used as a solvent, a cleansing agent and in paint manufacture.

5. Kerosene—Boiling range 140–290°C. This fraction includes domestic paraffin oil (140–250°C) and heavier slow-burning fractions used as illuminants in railways (signal oil) and in lighthouses.

6. Gas oil—Boiling range from 180°C and leaving a residue of carbon at 350°C. This is used as a carburetting oil in the gas industry and also in diesel engines.

7. Fuel oils—Boiling above 200°C. Beyond this point vacuum distillation is necessary to avoid cracking of these heavy oils with the formation of lighter oil and carbon residue.

8. Light fuel oil—Boiling above 200°C.

9. Heavy fuel oil—Boiling above 250°C

These are fractions of interest to furnace operators—for use in ships, land boilers, metallurgical furnaces, etc.

10. The residue from these stages is becoming very thick and yields on further treatment, if paraffinic, wax (paraffin wax), mineral jelly ("Vaseline"), and lubricating oils and greases or, if asphaltic, bitumens and lubricants.

Early *thermal cracking* techniques were developed about a century ago to use up superfluous heavy oils in the production of light paraffins for illuminants. Very brief heating at 400–500°C caused some charring but had the desired effect. Higher temperatures produced more gaseous products or polymerization to aromatics if above 700°C.

Today the accent is on producing gasoline rather than kerosene, but future developments in engine design may well shift the demand on to diesel fuel (gas oil) for internal combustion and jet engines.

The modern techniques are by *catalytic cracking* which gives greater control over the reactions occurring. The product can be controlled through temperature, pressure, catalyst and time of contact. Catalysts used in this stage were originally specially prepared bentonite clays but now tiny granules of synthetic alumino-silicates are employed and fluidized bed techniques are used to put the process on a continuous production basis. The cracking temperature is about 500°C. Poisoned catalyst is continually revived by burning off the char as it is passed round a regeneration cycle.

Further treatments may be applied to improve various fractions. *Reforming* is either a thermal or a catalytic treatment to improve straight-run gasoline or convert naphtha to gasoline. Catalysts used include platinum and oxides of aluminium, chromium and molybdenum. The aim is improved petrol with good anti-knock properties for high compression ratio engines, and this is achieved by increasing the proportion of branched hydro-carbons and aromatics.

Catalytic polymerization usually aims at combining olefine molecules in twos and threes to convert gaseous olefines to liquids for the gasoline fraction. Treatment with sulphuric or phosphoric acid is employed. *Alkylation* is a specialized combination of olefine and hydro-carbon molecules to give complex branched molecules characteristically anti-knock. Various acids act as catalysts for these reactions.

Any fraction may be purified by washing with reagents. Sulphuric acid removes unsaturated hydro-carbons and some sulphur compounds. Sodium hydroxide and lead dioxide react with sulphur compounds and remove sulphuric acid. Such processes are expensive and the cost of each operation must be justified by improved performance of the product. They are not likely to be applied widely to the heavy fuel oils though for metallurgical use a treatment to reduce sulphur would be worthy of consideration and would have an easily calculated economic value.

Oil shales provide only a tiny fraction of our oil requirements today, although the oil industry in Britain originated in the shale mines in the Lothians. Until 1962 this remained the only major shale oil producing area in the world with an annual output of 100,000 tons but production has now ceased. The oil produced was paraffinic and was treated in exactly the same way as petroleum after the initial distillation. There are vast reserves of oil in shales particularly in the U.S.A. and U.S.S.R.—greater than the known reserves of petroleum. It seems likely that these will ultimately be used.

Coal tar liquor is important to the steel industry which is usually situated near coalfields and carbonizing plants.

The tar may be distilled to the following fractions:

1. Light oils—b.p. up to 170°C yielding benzene and naphtha.
2. Carbolic oils—170–230°C, yielding phenols, naphthalene, etc.
3. Creosote oil—230–270°C, yielding motor-spirit, tars, acids.
4. Anthracene oil—270–320°C, yielding anthracene, etc.
5. Residual pitch which may be further distilled to yield hard pitch or pulverizable coal tar fuel.

While valuable chemicals are available in these products, the demand for them is limited and a large part of the tar is prepared into a series of standard coal-tar fuels designated C.T.F. 50, 100, 200, 250, 300 and 400. The number indicates the temperature in °F at which the blend is fluid enough for atomization (i.e. a maximum viscosity of 0·25 stoke).

C.T.F. 50 and 100 are blends of fractions (2), (3) and (4) above. C.T.F. 300 is a blend of (5) and (4) with varying amounts of (3) achieved by stopping distillation at an appropriate stage, while C.T.F. 200 and 250 are made by "oiling back" blend 300 with further additions of (3). C.T.F. 200 is the most popular blend. These fuels have high calorific value, low ash and sulphur content and give good luminous flames. They may not be mixed in tanks or pipelines with petroleum fuels, as tars are precipitated.

The final residue of hard pitch may be used as a pulverized fuel after comminution in an air-swept swinging hammer mill. It is obviously economically desirable that all these organic residues should be burned if possible.

The hydrogenation of coal to produce fuel oil is not at present important in this country and has been employed elsewhere only for making motor and aviation spirit and other light oils rather than those of metallurgical interest. The process involves auto-claving coal with hydrogen at about 500°C and 200 atmospheres pressure, and distilling the product. It can be applied also to other carbonaceous matter—e.g. oil, pitch, etc.

Special Tests on Liquid Fuels

Flash point is the minimum temperature at which the oil will catch fire if exposed to a naked flame. Minimum flash points are stipulated by law for various grades of oil (> 80°C for fuel oil), and values are determined in a standard apparatus (e.g. Pensky–Martin) used in a standard manner. The test is primarily carried out as a safety precaution, but would also indicate deviations from specification.

Viscosity is also determined in standard equipment as a check on specification as regular behaviour in pipelines and burners is desirable. Viscosity varies logarithmically with temperature and would best be determined over a range of temperature. The Redwood Viscometers (I and II) comprise an oil cup in a water jacket which can be temperature controlled. When conditions are stabilized the oil is allowed to flow through a standard orifice in the base of the cup and the viscosity is expressed as the number of seconds for the first 50 ml to flow. When the efflux time in I exceeds 2000 sec viscometer II is used, in which the orifice is wider and the time is reduced by a factor of 10. Conversion of Redwood (I) seconds to centistokes is by a factor of about 0·25 which is not, however, constant over the range. Viscosities of fuel oils lie in the range 250–7000 Redwood (I) seconds at 100°F (37·8°C). More viscous fractions are cracked to lighter components. Oil can be pumped at a viscosity of about 3000 sec but must be warmed to 100 sec before it can be atomized. Steam for atomizing provides some of the necessary heat. Viscosity measurements over a range of temperatures are more informative than the earlier "pour point" test which determined directly the temperature below which an oil would not flow out of a standard size of test tube.

Other determinations of special interest include density and its variations with temperature to provide an accurate conversion factor for volume to mass, especially if liquid fuels are metered and bought by volume. Specific heat data are also useful for assessing heat requirements when warming oil to atomizing temperature. Moisture content is determined, usually on residual fuel oils, by distillation assisted by an addition of volatile petroleum spirit with which water is immiscible. The volume of the water in the combined condensate is measured directly. Light oils and lubricants are subjected to other tests such as the measurement of carbon residue when all volatiles are burned away, the determination of asphaltenes, or the assessment of distillation characteristics. These need not be considered further here.

Properties and Uses

The properties of a selection of liquid fuels are arranged for comparison in Table 5. It will be observed that the sulphur is concentrated in the heavier fractions of the petroleum oils. For many purposes this is unimportant. It is, of course, an advantage to the user of the light fractions. In steelmaking, however, the high sulphur content of fuel oil is probably its most objectionable feature and sets oil at a disadvantage compared with coal-tar fuels. Ash also concentrates in the heavy fractions but not so far as to be of significance in furnace work.

TABLE 5
Properties of Some Liquid Fuels

	Motor gasoline	Motor benzole	Kerosene	Diesel (gas) oil	Light fuel oil	Heavy fuel oil	C.T.F. 200	Bituminous coal (for comparison)
Carbon %	85·5	91·7	86·3	86·3	86·2	86·2	90·0	80·0
Hydrogen %	14·4	8·0	13·6	12·7	12·3	11·8	6·0	5·5
Sulphur %	0·1	0·3	0·1	1·0	1·5	2·0	0·4	1
Nitrogen %	—	—	—	—	—	—	1·2	1·5
Oxygen %	—	—	—	—	—	—	2·5	7
								5 (Ash)
Specific gravity (15°C)	0·73	0·88	0·79	0·87	0·89	0·95	1·1	1·25
Kinematic viscosity:								
20°C cS	0·75	0·72	1·6	5·0	50	1200	1500	—
Redwood I sec	—	—	—	—	200	5000	6000	—
100°C cS	—	—	0·6	1·2	3·5	20	18	—
Redwood I sec	—	—	—	—	30	90	80	—
Boiling range °C	40/185	80/170	140/280	180/—	200/—	250/—	200/—	—
Residue at 350°C %	—	—	—	15	50	60	60	—
Flash point °C	−40	−40	39	75	80	110	65	—
Calorific value (net):								
cal/g	10,450	9600	10,400	10,300	10,100	9900	9000	7750
kJ/kg	43,740	40,200	43,500	43,100	42,300	41,450	37,700	32,450

The calorific values of all fractions are similar and oils have rather a higher calorific value than coal-tar fuel, and a considerably higher value than coal, especially if translated to a (bulk) volume basis. These differences are due to the higher oxygen and ash contents of tar and coal.

It will be apparent that the viscosities of heavy oils vary over a tremendous range and it will be obvious that for convenience in furnace operation a supply of fuel of an almost constant viscosity is highly desirable. Some adjustment can be made through temperature but this should not be necessary except under unusual circumstances. Fuel oil is most commonly specified by viscosity at one or more temperatures. There are British Standards Specifications for fuel oils which can be ordered by naming the appropriate standard grade. Alternatively, closer specifications can be demanded and can usually be met by blending. In other countries similar standard grades are specified also. Major differences might arise in the oils offered by different companies if these originated in different oilfields. Thus, one oil might be essentially paraffinic and another asphaltic, and these might be expected to behave slightly differently in the furnace especially in the last stages of combustion. They might also differ somewhat in appearance, in ash content, and in the nature of the ash. Differences of this kind are of greater importance when the oil is being used as fuel for engines.

The uses of oils are as varied as those of gas and include domestic heating and lighting, steelmaking, all kinds of engines, the generation of electricity, and the production of gas. Oil is also used as a raw material in the chemical industries. It can be almost as clean as gas in use but spillage and leaks of thick oil and tar are unpleasant features of many plants. The need for steam for warming the storage tanks and pipelines and for injection through the burners involves the use of space and money and the delivery of oil in quite small and very frequent batches by road tanker necessitates some careful organization and numerous checks on quality.

Oil can be burned to very intense hot flames, especially if used with oxygen-enriched air. In this respect only the very rich gases can be expected to do as well. On the other hand, oil flames are luminous and radiate heat very effectively while rich gases burn to clear non luminous flames and tar or oil must be added to provide luminosity. Coal-tar fuels give flames which have particularly good luminosity and heat transfer rates.

The use of oil has increased in recent years and it has apparently displaced coal and gas from some of their traditional preserves. In fact the supply of oil has expanded more rapidly than that of coal to meet the rapidly increasing demand for fuel of any kind, but it may be worth observing that while oil is replacing gas in, say, steel plant, it has itself virtually disappeared as an illuminant, having been replaced first by gas and then by electricity. The advantages of oil in steel plant were primarily that it was available at a critical period when coal was scarce and secondly that oil benefited from the post-war developments in steelworks fuel technology and now stands in a very strong position compared with gas. Higher production rates are being obtained (partly due to other causes such as the use of tonnage oxygen) and much of the industry is geared to its use. The advantages of oil over gas or coal in other fields is not so substantial, however, and at the level of domestic central heating, for example, the significant differences between gas, oil and coke-fired appliances are likely to be commercial rather than technological or economic.

Oil Fuel for Transport

About 25–30 per cent of petroleum products is used for locomotive purposes—about 5 per cent in ships, 17 per cent on roads, 4 per cent by aircraft and the remainder by railways, by agriculture and in other ways.

Heavy fuel oil is used in large ships for raising steam which then drives the ship through turbines. This fuel is used like any industrial fuel in furnaces. The most important quality is consistency in the viscosity and it is necessary that standardization on a world-wide scale makes re-fuelling with the same grade possible at any port.

Many ships are now powered with diesel or compression ignition engines. These are more compact, cleaner and more efficient than steam turbine engines but the fuel required, being more refined is more expensive. Diesel oils come from the heavy end of the straight run distillate range or may be obtained by cracking of still heavier fractions, and consist of paraffins in the range C_{13}–C_{20}. In principle any liquid fuel with appropriate characteristics could be used, such as creosote oils, and mixing and blending either for better performance or lower price is commonly practiced.

Diesel oils are distinguished mainly by their ignition properties. Quite small variations in general properties such as S.G., viscosity, flash point and distillation characteristics can mask large differences in ignition characteristics which determine whether the oil will ignite and burn quickly or slowly when injected into the cylinder of the engine. The oil must be able to ignite at a low enough temperature that it can be used even in a cold engine and the moment at which it ignites must not be delayed too long beyond the moment of its introduction into the cylinder. The permissible delay varies with the design of the engine. Large, slow revving engines with speeds below 300 r.p.m. can use oil which ignites sluggishly but high revving engines clearly require that the reactions are completed in a shorter time, so delay of ignition cannot be tolerated.

The "ignitability" of an oil is expressed as its "cetane number". Cetane ($C_{16}H_{34}$) is the fastest igniting hydrocarbon in the range and it is allotted the cetane number 100, while an aromatic hydrocarbon, methyl naphthalene, is arbitrarily rated zero on the cetane scale. An oil is compared with mixtures of these, as it behaves in a standard engine, and rated by the percentage of cetane in the mixture which it most nearly matches. In practice secondary standards are used or a quicker, cheaper indication can be obtained through the aniline point test which effectively determines the proportion of simple paraffins present in the oil.

Oils with cetane numbers about 20 can be used in large low revving engines as on ships but for use in a motor car for example where, say, 4000 r.p.m. would be required, an oil would be needed with a cetane number of about 60. Diesel oils should be very low in water and ash and also in asphalts which tend to leave carbonaceous residues in the engine. Low revving engines can be run on oils with more and harder asphalts than can high revving engines because there is more time available in each cycle in which the residues can burn out. Diesel oil made from cracked residual oils are more likely to be rather high in ash and asphalt and also have low cetane rating but it would also be cheaper and be quite suitable for large slow engines. Apart from their use in ships and motor cars, diesel engines are used in railways, often coupled with electricity generators, and in heavy road vehicles. They are also common on stationary plant including generating stations for meeting peak demand for electricity.

Gasoline is a much lighter cut of the paraffins ranging between C_4 and C_9. These may be doped with benzene or with alcohols. The former gives a denser and the latter a less dense fuel than average fuel which becomes important when they are being marketted by volume. Both benzene and alcohol have been used alone in internal combustion engines but neither is so used today.

The important characteristics distinguishing gasolines are their ignition and combustion behaviours. Internal combustion engines work most efficiently at high compression ratios but this can lead to spontaneous ignition (as in the diesel engine) before the spark can be fired. This is not desirable and is manifest as "pinking" or "knocking". Some paraffins are more prone to this than others and in general the long chain hydrocarbons are most likely to exhibit the phenomenon, while multiply branched hydrocarbons show it least. The proportions of straight and branched molecules varies with the source of the oil but in most cases the straight run distillate has insufficient of the branched molecules and these must be augmented by blending in catalytically reformed naphtha rich in branched isomers. This raises the price of the fuel but enables the engine to be run more efficiently.

This quality of gasoline is expressed in terms of its "octane number". This is rather like the cetane number. Pure iso-octane is designated as having an octane number 100 while normal heptane is assigned the rating zero.

```
     H       H
    HCH     HCH
 H   |   H   |   H          H  H  H  H  H  H
HC—C—C—C—CH               HC—C—C—C—C—CH
 H   |   H  H  H            H  H  H  H  H  H
    HCH
     H
   iso-octane                   n-hexane
```

These are the most completely branched and the least branched, and the most and least anti-knock compounds in the range. Gasolines are compared in their behaviour in a standard engine with mixtures of these and assigned as their anti-knock rating the number indicating the percentage of iso-octane in the mixture which the test sample most nearly resembles. In practice secondary standards are used. If a high enough octane rating cannot be obtained by blending hydrocarbons other additives may be incorporated. The most common of these is tetra-ethyl lead along with ethylene di-chloride to prevent the lead depositing on the valves. Using this the octane number can be pushed up beyond 100 but legislation limits the concentration of lead which may be used to 0·84 g/l. and the modern trend is toward the re-designing of engines to use lower octane fuels with a view to eliminating lead if possible. Some rather expensive cars are now equipped with such engines.

Gasolines must be low in gums or gum forming compounds which may leave behind unburned residues. Water must also be absent to avoid corrosion of tanks. A suitable proportion of the lower boiling paraffins should be present to facilitate cold starting.

Aviation gasoline has a rather lower boiling range and a higher octane rating than motor fuel. It should be even freer of gums than motor fuel and is more costly. The demand for this fuel is steadily declining.

Most aircraft are now fitted with jet engines which use kerosene which is much cheaper than gasoline. Some engines require a mixture of the two. This is not favoured because of the fire hazard but the kerosene used comes from the lower end of the kerosene range of hydrocarbons C_{10}–C_{16}. It is preferred with short branched molecules for rapid complete combustion and its water content should be as low as possible to avoid formation of ice at high altitudes.

Vaporizing oil is a light kerosene formerly used in tractors in internal combustion engines. These had to be started on petrol but once started and warmed the engines would run on the cheaper fuel.

8

Electrical Energy

ELECTRICAL energy is not a fuel but is a substitute for fuel and replaces it (or displaces it) in many industrial applications.

Sources of Power

Electricity can be generated from several sources of energy.

1. By conversion of chemical energy in coal, oil, peat or other conventional fuels into heat by burning, into mechanical energy by steam-raising, and then to electrical energy by using a turbine and dynamo.

2. By conversion of the potential energy in water in elevated reservoirs to kinetic energy in raceways and then to mechanical and electrical energy using turbines and dynamos.

3. By similar conversion of the kinetic energy in wind.

4. By conversion of nuclear energy via heat and steam.

5. By the use of electrolytic cells and fuel cells in which chemical energy can be converted directly to electrical energy.

Method (1) predominates in Britain and in most of the highly industrialized countries where coal is plentiful. The most modern boiler plant uses low-grade pulverized coal to produce steam at up to 850°C and 150 bar pressure. Steaming rates are very high at up to nearly 2 million kg/hr equivalent to about 200 tons of coal per hour, and the efficiency of energy conversion can be as high as 88 per cent at this stage. The modern furnaces do not use banks of water tubes athwart the combustion gas stream and relying on convection for heat transfer. These have been replaced by a jacket of water tubes placed round the walls of the combustion chamber and heated mainly by radiation from the luminous flame of the pulverized coal burned to a very high temperature with preheated air. The steam is expanded through the stages of a large turbine mounted on the same axis as the generator, the low-pressure side being almost a vacuum imposed by the condenser. The overall efficiency rises to about 38 per cent in best practice but the average is much lower and must remain so until the oldest plants working at under 20 per cent efficiency are replaced. The most efficient new plants, called base load plants, are designed to run at over 80 per cent capacity for 24 hours per day, peak loading of the grid being built up using the older units.

In Britain, water power has been developed mainly in Scotland but the total energy available is not great and its further exploitation can be increased only at rapidly increasing cost both in money and in "amenity". Most of the easy sites have been used up. The capital cost of a hydro-electric power scheme is high but the on-going costs are moderate and the water is, of course, delivered free of charge. Unfortunately the rate of delivery is not under control and varies with the seasons. Usually power is drawn intermittently to meet peak load requirements. Other more mountainous countries can produce a much higher proportion of their total energy requirements from water power but many of these also suffer either seasonal or occasional shortages of power due to either drought or frost. The water power from large continental rivers is now providing energy to several of the under-developed countries of Africa and Asia as it has already done in Russia. In some of these cases the "head" of water is not great but the volume of water available is very large. River schemes can be usefully adapted to provide controlled supplies of water as well as power for agriculture or industry. In Britain some of the hydro-electric schemes are used for "pumped storage" of water. The principle is that low cost energy from efficient base load coal, oil or nuclear stations should be used at night, when demand is otherwise insufficient to keep them going, to pump water into high reservoirs from which it can be drained, through generators at times of high demand when the base load stations are unable to cope with consumers' needs. The same motor/generators can be used both for pumping and for generating—working in opposite directions. The economics of such an arrangement depends on detailed costs in the light of local circumstances but pumped storage does provide one means of matching supply to a fluctuating load, which becomes acute when the supply comes from very large production units which, while very efficient while running, are too costly to be allowed to stand idle, even overnight.

Tidal power is also inexhaustible but costs a lot of money to harness and can be exploited only at very favourable sites. The French are ahead in this field with schemes in Mauritius and in France. In times of inflation these high capital cost ventures look to be excellent investments, ensuring that there is a good supply of apparently low cost energy in the years ahead.

Nuclear Energy

The production of electricity from nuclear energy is still in its infancy although many countries now possess large nuclear power stations which are operating successfully. The technology involved can be expected to make considerable advances in the next few decades. The basic principles are well understood but the complexity of the engineering and metallurgical problems involved is very high and the price of a single failure could be enormous.

In all cases, heat produced in a nuclear reactor must be used to raise steam which is then used to drive turbines coupled to generators in a conventional manner. Because of the properties of the materials in the reactor it is not possible to raise the temperature there above about 550°C which is much lower than the flame temperature in a conventional boiler fired with coal or oil. The temperature of the steam produced is, consequently, comparatively low and the efficiency of conversion from thermal to electrical energy is also low. A dual pressure steam heating cycle is used in the heat exchanger to improve the heat exchange efficiency between the reactor and the turbines but the low steam temperature available nevertheless puts a severe limitation on the overall efficiency of conversion.

All commercial reactors use uranium as the primary fuel but thorium could be used and plutonium will doubtless soon be used for enrichment once the fast reactor technique becomes established. The uranium may be natural, mainly U^{238}, or it may be enriched with respect to the U^{235} isotope of which there is only 0·7 per cent in the element as mined. Neutrons, which arise spontaneously in radioactive materials under the internal bombardment of their own radiations, can cause the disintegration of atoms of U^{235} into two smaller atoms, with the release of one or more further neutrons and other radiations, the energies of which are largely convertible to heat. The mass of the parts formed in any disintegration is less than that of the original uranium atom by a small amount δm which is related to the change in the internal bonding energies δE by Einstein's equation $\delta E = \delta m \; c^2$, where c is the velocity of light. The neutrons released may escape the system if it is small, or they may be absorbed by U^{238} atoms to form a new isotope U^{239}, or they may collide with U^{235} atoms to cause further disintegration and more new neutrons. If, on average, one neutron from each disintegration brings about one other disintegration, a chain reaction has been established and the process is self-sustaining. The energy released can be withdrawn as heat and put to useful work. Actual operation of an atomic "pile" depends on conditions being arranged so that a much higher proportion of neutrons brings about disintegrations than will ever occur in a large piece of natural uranium because the U^{238} absorbs too many of them. The effective proportion depends on the geometric design, the neutron energies and the kinds of atoms present. It can be increased by using a more spherical form so reducing the escapes per unit volume, but greater increases are effected by increasing the proportion of U^{235} (enriched fuel) or by slowing up the neutrons by passing their radiations through materials called "moderators", light elements carbon, beryllium or deuterium, which slow up the neutrons to so-called "thermal" energies. They are then more likely to react with U^{235} and less likely to be absorbed by U^{238}. The core of a so-called "slow" or thermal reactor is a lattice made up of a large mass of moderator permeated by rods of fuel which may or may not be slightly enriched. The fuel itself must be kept free of impurities which would absorb neutrons without producing heat. Disintegrations of course produce such impurities and the fuel must be purified of them periodically—a difficult and expensive chemical operation, especially considering the highly radioactive nature of all the materials involved. When the proportion of U^{235} is high—over about 25 per cent—a "fast" reactor may be operated with no moderator beyond the uranium itself. This type of plant would be quite compact and it could produce heat just as fast as it could be withdrawn from the pile. If the excess of neutrons or those escaping from the core can be captured by U^{238}, some plutonium can be produced which can be extracted chemically and used as the fuel in another fast reactor of suitable design, since plutonium is similar to U^{235} in that it is readily disintegrated by neutron bombardment. The reactor producing the plutonium is called a "breeder" reactor.

At the present time all commercial reactors are "thermal" reactors but "fast" or "breeder" reactors are working, that at Dounraey producing about 250 MW of electricity, but still looked on as a large-scale experiment. A sphere of natural uranium surrounds a small core made up of fuel elements of enriched uranium containing about 25 per cent of U^{235}. The fuel elements are very narrow "pins" with a high specific surface for good heat transfer. Control is effected by moving these fuel elements in and out of the core. These become contaminated with fission products and must be removed for purification and re-fabrication at regular intervals. This is very costly. To achieve economic working conditions the heat abstraction rate must be very high. This is done by using as the heat transfer medium a liquid metal–sodium or a sodium potassium alloy. Because this sodium could become radio-active it is not allowed to circulate outside the radiation shield but exchanges its heat with a secondary sodium circuit which in turn passes the heat over to a steam circuit which serves the turbines. The core of this reactor comes nearer to "atomic bomb" conditions than any other reactor and this along with the extremely reactive nature of the sodium makes this kind of reactor particularly dangerous in the sense that some kinds of "accidents" could be absolutely disastrous. It is therefore essential that every possible breakdown shall be anticipated and fail safe devices incorporated which will disperse fuel, quench fires and contain the products of explosions. The development of equipment of this kind is understandably slow but the advantages of success will be enormous because the "fertile" uranium in the "blanket" around the core captures neutrons and is converted into Pu^{239} which can be used as enrichment for fuel elements in another reactor. Ultimately the costly production of U^{235} will be unnecessary. Whereas a ton of natural uranium provides energy equivalent to about 10^4 tons of coal a ton of plutonium provides as much as 3×10^6 tons of coal and in use can generate yet more from natural uranium. At that rate about a hundred tons of plutonium would meet all Britain's energy needs for a year—along with a similar amount of uranium as the basic raw material.

Thermal reactors are comparatively simple but larger and less efficient. The fuel may be natural or slightly enriched uranium and it must be surrounded by a "moderator" as explained earlier. The variety of different types of reactor which have been suggested, lies in differences in (1) the moderator used and (2) the heat exchange medium. The first major successful design, called "Magnox", used CO_2 as the heat exchange medium. It is therefore a gas cooled reactor. Fuel elements are of natural uranium enclosed in "cans" made of a magnesium alloy, the name of which, magnox, was transferred to the system. The can is necessary to contain the fission products and prevent contamination of the CO_2 with radio-active products. The moderator is graphite of high purity—a very large block of graphite penetrated by cylindrical channels through which fuel rods assembled from canned fuel elements, and control rods made of steel can be moved. The CO_2 remains radio-actively clean and is used directly for steam raising. It was necessary to enclose the whole core in a pressure vessel to allow enough CO_2 to circulate to abstract heat at a satisfactory rate. These reactors operate satisfactorily at about 500 MW output.

Improvements in design have led to the advanced gas cooled reactor. These were proposed first in 1965 but the state of the art is such that ten years later none have been commissioned although several in Britain are very nearly completed. The fuel in these is UO_2 which overcomes problems arising from the behaviour of metallic uranium under irradiation. The cans are made of stainless steel which can be taken to a higher temperature than magnox. A more efficient heat withdrawal is achieved using a much more sophisticated shape of can. The heat exchange medium is again CO_2.

Other designs use either water or deuteria as the moderator. Boiling water reactors use enriched UO_2 fuel canned in zirconium and immersed in water which acts as moderator and heat transfer medium. The steam produced passes its heat in a primary exchanger to a second steam circuit which operates on the turbines. An alternative technique is to use pressurized water as the moderator and heat transfer medium—the water not being allowed to boil in the reactor but the hot water is circulated through a heat exchanger in which water in the secondary circuit is evaporated. These designs are more compact than gas cooled reactors. They appear to be inherently less safe but a large number of stations have been operating successfully in U.S.A. and elsewhere for several years.

In these designs the use of heavy water D_2O instead of H_2O would be an advantage, except in cost, because D_2O does not absorb neutrons as does H_2O and some successful designs do use that isotope. Other designs have been suggested in which the uranium is in solution and the solution is circulated as the heat transfer medium. Clearly in all cases the technological difficulty is great. Corrosion must be prevented for an indefinite period and metallurgical creep and fatigue must not occur. Running repairs are difficult to put into effect and where serious defects occur the abandonment of entire plants could become unavoidable. The capital cost of such losses would bear heavily on the actual cost of the energy produced.

Another cost incurred in the production of electricity from uranium is the cost of disposal of the dangerous radio-active fission products. Indeed there is the problem of what to do with these substances—how to dispose of them in such a way as can never cause harm to future generations. Clearly some sort of international agreement on where the increasing amounts of this rubbish may be dumped, is badly needed.

Control of the output of a reactor is effected by means of control rods. In slow reactors these are made of steel and act by absorbing a high proportion of the neutrons they encounter. In fast reactors the rods are fuel elements and the activity is reduced or stopped by reducing the mass of the fuel below a critical value necessary for maintenance of the chain reaction. The heat developed is removed from the pile by the circulation of a suitable fluid through a heat exchanger designed into the pile and used to make steam in another exchanger external to the pile. Carbon dioxide and liquid sodium have been used as heat transfer media, the latter, having the higher heat capacity per unit volume, being used in the more compact fast reactor.

There are still many metallurgical problems only partly solved pertinent to the operation of the reactors and research is being carried out into the best forms of both the moderators and the fuel elements and to find whether pure metals, alloys or refractory compounds are likely to be the most suitable for operation at higher temperatures. At present the physical properties of the materials used limit the operating temperature to about 500°C, which puts the generation of power from atomic energy at a severe thermodynamic disadvantage compared with conventional methods.

Fuel cells are still in an early stage of development. They could be used to produce small amounts of direct current in isolated sites but their commercial development for the large-scale generation of electricity seems a long way off. The principle is that the free energy of a reaction, which would usually be, in effect, the oxidation of a fuel, is tapped off as electrical energy when the reaction is made to proceed as two partial reactions at conducting surfaces in an electrolytic cell. A necessary transfer of electrons is obliged to take place by their travelling along an electrically conducting path joining these electrodes outside the cell. The simplest system would be the oxidation of hydrogen. The conditions are not simple. Hydrogen and oxygen must be delivered separately into the cell each being passed through a porous sintered metal electrode. The hydrogen must ionize, and the electrons liberated go to the other electrode where they enable oxygen to combine with water molecules to form hydroxyl ions. Within the electrolyte the hydrogen or hydroxonium ions combine with the hydroxyl ions to form water. The cell conditions necessary are not easily arranged—an alkaline solution at high temperature and pressure and critical gas flow rates. The output from each cell is available only at about one volt but the current density on the electrodes can be quite high at about 10^4 A/m^2. In principle batteries of such cells could be assembled to provide electricity at suitable voltages for at least local distribution. To be really useful it would be necessary to develop cells which would work on cheaper fuels and preferably on solid fuel such as powdered coal. This might be possible using fluidization techniques. The great advantage that would be gained is that the efficiency of conversion of chemical to electrical energy is theoretically 100 per cent because it does not involve heat at any stage. The Carnot cycle limitation does not apply. It could come to pass, however, that the hydrogen fuel cell does eventually play a major role in fuel technology. If nuclear electricity became cheap it might be more convenient to transmit that energy as hydrogen produced by electrolysis if its efficient reconversion to electricity could easily be effected by fuel cell near to the point of use.

Production

Electricity is generated in an alternator by rotating an assembly of conducting coils in a magnetic field. The coils, mounted on a shaft called a rotor, comprise the armature. The magnetic field is formed by a series of electromagnets energized by direct current supplied from an external source. The coils terminate in slip rings from which current is drawn through carbon brushes into the load circuit.

In a very simple case there might be two magnetic poles, North and South, and a single coil of wire. If this were rotated once in the magnetic field with constant angular velocity, the e.m.f. developed on open circuit (or the current through a constant load) would rise to a maximum, fall to zero and then to a minimum and then rise to the original value as the coil returned to its initial position, the graph of voltage against time t describing a sine curve, $V = V_m \sin \omega t$ (Fig. 4(a)). This would correspond to one electrical cycle.

If there were four poles, North, South, North, South, one revolution of the coil would complete two electrical cycles, and so on. Further, if there were two pairs of equally spaced poles, two coils could be used, set at 90° to each other, and these would augment each other insofar as twice the current could be supplied. In such a case there would be a lot of space on the rotor not occupied by coils and other pairs of coils could be accommodated. If two more pairs were built on to the rotor at 30°, 120°, 60° and 150°, the original pair being at 0° and 90°, a three-phase supply

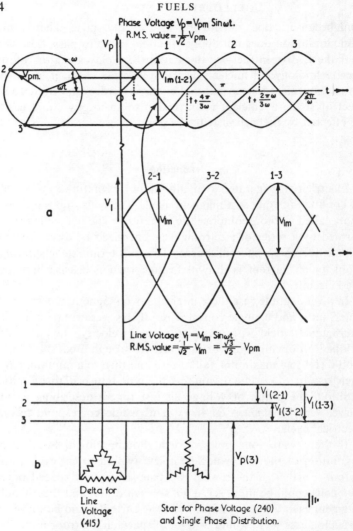

Fig. 4. Three-phase production and distribution of electricity.
(*a*) shows the relationship between the phases and that between
phase and line voltages, (*b*) shows how delta and star connections
are made at distribution points.

would be obtained. Each set of coils would have its own pair of slip rings and, assuming the connections to these were made in a symmetrical manner, the timing of the electrical cycles of the three phases would be different from one to the next by one-third of a cycle. In electrical parlance the phases are displaced by 120°. See Fig. 4(a).

In practice three-phase generation is the rule. The number of poles may be much higher than four and the magnetic field usually rotates with the rotor while the conducting coils are built into the stator. This simplifies the slip-ring problems. There are many detailed variations on this scheme but these need not be considered here.

The alternator is driven at such a speed as will give exactly 50 electrical cycles per second, and this is the frequency of the supply, 50 Hz (in the United Kingdom). The efficiency of conversion of mechanical energy to electrical energy is a function of alternator design and engineering and metallurgical skill in achieving what has been designed. The clearance between the rotor and the stator must be very small and constant. The coils are wound not on a solid former but on an armature core built up of hundreds of thin plates of iron insulated from one another with thin paper so that the coils and not the core carry the current. Overall conversion efficiency has increased in recent years from about 20 per cent pre-war to about 38 per cent now. Generators have increased in size at the same time from about 30 MW to 600 MW capacity, with 1300 MW sets now being built.

Direct current can be produced in a similar device but the slip-rings are segmented so that current can be drawn only when it is flowing in one direction. This part is known as the commutator. Change-over from a.c. to d.c. or from one frequency to another can be effected by motor-generator sets in which the supply available is used to drive a motor coupled to a generator which has characteristics suitable for producing the kind of supply required.

Distribution

Electricity for general distribution is generated at about 12,000–20,000 V a.c. For distribution this is transformed up to one of the standard transmission voltages—400 kV on the national super-grid, 132 kV on the nation grid, and back down to 33 kV or 11 kV for regional distribution (in the United Kingdom). Tappings from the grid system are transformed down by stages for local distribution and voltages of 240 and 415 are ultimately available in the home and the factory for general purposes.

The three phases must be kept in balance and except for domestic users three-phase supply is the rule rather than the exception, and the consumer is expected to load them equally and keep them in phase. The phases are interconnected for transmission either by Delta (mesh) or Star (Y) connections as shown in Fig, 4(b). In each of these any two wires are continuously acting as the return for the third, but domestic distribution uses a fourth wire connected at the pole of the Star at the sub-station transformer, essentially earth potential. The normal voltage between line and this "neutral" wire is 240 (phase voltage), while that between phases in the Delta arrangement is 415 (line voltage). This provides a very good reason for maintaining single-phase supply in homes. Balancing the phases in domestic circuits is done statistically.

240 V single-phase supply is used for all domestic heating, lighting and power, and in laboratories most equipment up to 3 kW can be run off this supply. Industrial equipment involving heavy motors—lathes and other tools for example—would require three-phase, 415 V supply and be designed to use it. Industrial electric furnaces would require to be supplied off the three-phases, but usually at lower voltages so that the current and therefore the heating effect would be as high as practicable. Transforming down to low voltages is done near the furnace so that the heavy conductors required for the high currents will be short.

It will be realized that (at equal energy flow rate) a rise in voltage is accompanied by a proportional fall in current and therefore in transmission losses. Low-voltage high-current trans-mission requires bus bar connections to be very heavy, otherwise they will melt. The choice of 240 V for domestic supply in Britain is a compromise between lethal voltages and incendiary amperages.

Considering a single phase, if a voltage is applied across a load which has neither inductance nor capacitance the current which flows will be proportional to voltage at any instant and will therefore vary sinusoidally in phase with voltage. At any instant Ohm's Law is being obeyed and Watts = Volts × Amps. Inte-grating over a whole cycle the effective volts is the root mean square (RMS) value or 0·707 $(1/\sqrt{2})$ of the maximum or peak volts, and the effective amps during the cycle is 0·707 of the maximum amps so that the watts dissipated equals $\frac{1}{2}$ × Peak Volts and Peak Amps.

Alternating current meters measure RMS values of volts and amps, so that in non-inductive circuits the formula can be used that Watts = Volts × Amps (the indicated values).

In inductive circuits, however, the current continues to vary sinusoidally, but not simultaneously with voltage—a little behind, a fraction of a cycle expressed as so many degrees lag (where a degree is 1/360 cycle). Volts and amps are said to be out of phase and where θ is the angle of lag, power becomes Volts × Amps × Cos θ.

If a circuit has capacitance the amps may be in advance of volts by an angle θ with the same effect that power = Volts × Amps × Cos θ is less than it would be in the circuit with no in-ductance or capacitance. Cos θ is known as the power factor (p.f.) and depends on inductance and capacitance taken together.

As power is measured and charged as KVA and energy is usefully dissipated as KVA × p.f., it is highly desirable that the power factor be kept as high as possible and where necessary capacitance is introduced deliberately to advance the current so that the inductance in the load will retard it only to the optimum position—in phase with voltage. Most large electric furnaces require such capacitance banks to be installed.

Big industrial consumers frequently produce at least a part of their own electricity. This may be to effect economic use of waste heat or excess gas from their major operations. It is also of some benefit in ensuring safety in the case of grid failures and in making some provision for peak loading periods. The tariff rates to industrial consumers vary with the type of equipment operated and in some cases are proportional to the rate at which power could be consumed on the grounds that so much generating capacity has to be available at all times. In cases like this, e.g. electric melting furnaces, the use of electrical energy can be justified only if the operation is continuous, round the clock. Lower rates may be available for power consumed overnight.

Electricity is used in three principal types of furnace—resistance, arc and induction furnaces. Resistance furnaces are in most general use in many industries at low and moderate temperatures. They provide excellent uniform heating and essentially "muffle" conditions without the disadvantage of a refractory screen interposed between the heating elements and the stock. Arc furnaces are used mainly by the steelmakers and particularly by manufacturers of expensive high-alloy steels. Induction furnaces are used for melting special steels and other alloys such as brasses and nickel alloys, and can be applied to very high-class work such as vacuum melting. Electricity is also used in a few specialized metallurgical extraction techniques. Generally in Britain the cost of electricity is too high to justify its replacing fuel except where some distinct technological advantage is to be gained, as in the cases of the high-alloy steels mentioned where only electric furnaces can be operated under the neutral or reducing conditions necessary for attaining the desired quality. Electricity is very easily controlled: it makes very high temperatures very readily attainable; and it is very clean to use particularly insofar as it produces neither sulphur nor ash at the point of use. There is at present a clear trend toward its extended use in steelmaking. The normal size of arc furnaces has in recent years increased and whole melting shops have been converted from gas-fired to electric arc furnaces for the manufacture of a wider range of steels than has hitherto been made in these furnaces. The economic explanation of these developments is that, especially where steel is being produced from cold metal charges, the melting down time is much shorter in electric arc furnaces than in fuel-fired furnaces and the advantages of increased productivity outweigh the higher price of electrical energy.

9

Trends in Fuel Utilization

IN Fig. 5 an attempt has been made to present the present-day utilization pattern of energy in the United Kingdom in diagramatic form. The figures are in millions of coal ton equivalents to allow a direct comparison to be made of the energy used at each point but the equivalence of different fuels is not unambiguous. Coal used for coke manufacture, for example, would have a higher calorific value than most of the coal sent to power stations but no particular account has been taken of this difference. Energy provided as electricity is clearly derived from about four times as much coal and oil energy as the amount shown against the electricity produced. The steel industry has been given fairly detailed treatment in the diagram but in other sectors the information readily available is much less complete. The overall picture is, however, reasonably clear.

Rather less than half the total energy requirements (44 per cent) of the country was met in 1974 with coal while oil contributed 35 per cent, natural gas 20 per cent and nuclear energy only about 2 per cent. The corresponding figures in 1960 were coal over 80 per cent and oil just under 20 per cent with negligibly small contributions from natural gas and nuclear fuel.

Today over half the coal used goes to power stations along with one-third of that part of the oil which is used for heating purposes. This is converted with an average efficiency of about 28 per cent into electrical energy equivalent to 27·9 tons of coal. 2·5 mtce of this energy—about 8 per cent of the total electrical energy

OIL

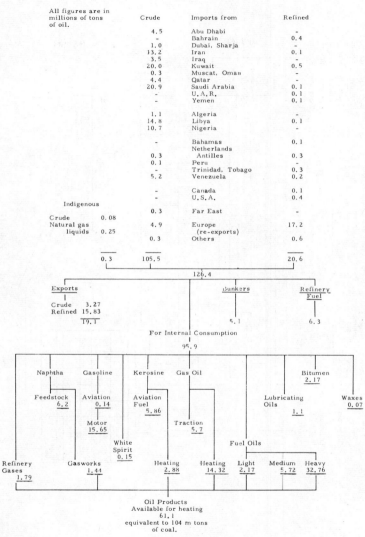

All figures are in millions of tons of oil.	Crude	Imports from	Refined
	4.5	Abu Dhabi	-
	-	Bahrain	0.4
	1.0	Dubai, Sharja	-
	13.2	Iran	0.1
	3.5	Iraq	-
	20.0	Kuwait	0.5
	0.3	Muscat, Oman	-
	4.4	Qatar	-
	20.9	Saudi Arabia	0.1
	-	U.A.R.	0.1
	-	Yemen	0.1
	1.1	Algeria	-
	14.8	Libya	0.1
	10.7	Nigeria	-
	-	Bahamas	0.1
	0.3	Netherlands Antilles	0.3
	0.1	Peru	-
	-	Trinidad, Tobago	0.3
	5.2	Venezuela	0.2
	-	Canada	0.1
	-	U.S.A.	0.4
	0.3	Far East	-
	4.9	Europe (re-exports)	17.2
	0.3	Others	0.6

Indigenous

Crude 0.08
Natural gas liquids 0.25

0.3 105.5 20.6

126.4

Exports

Crude 3.27
Refined 15.83

19.1

Bunkers 5.1

Refinery Fuel 6.3

For Internal Consumption

95.9

Naphtha Gasoline Kerosine Gas Oil Bitumen 2.17

Feedstock 6.2 Aviation 0.14 Aviation Fuel 5.86 Lubricating Oils 1.1 Waxes 0.07

Motor 15.65 Traction 5.7

White Spirit 0.15

Fuel Oils

Refinery Gases 1.79 Gasworks 1.44 Heating 2.88 Heating 14.32 Light 2.17 Medium 5.72 Heavy 32.76

Oil Products
Available for heating
61.1
equivalent to 104 m tons
of coal.

FIG. 5—PART 1

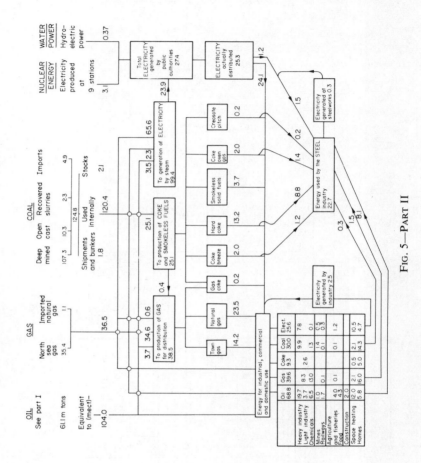

FIG. 5—PART II

produced—are dissipated in transmission so that the high cost of electricity per energy unit can well be understood and the need to use it efficiently should be appreciated.

About 20 per cent of the coal used goes to the manufacture of coke and smokeless fuels. About half of this energy goes to iron and steel works as coke and gas and the remainder to industrial and domestic purposes.

Of the oil imported to Britain, Fig. 5 shows that about 60 per cent is available for heating purposes, not counting 7 per cent consumed within the oil refineries. Over 25 per cent of the oil imported is used in transport—or 30 per cent if oil for ships bunkers is included. Motor fuel accounts for about 16 per cent. The remaining 6 or 7 per cent is largely used by the petrochemical industries as feed stock or as lubricants, bitumens and waxes.

Of the oil used for heating 30 per cent goes to power stations, a small amount to the production of gas—this is diminishing rapidly—and the rest goes directly to industrial and domestic consumers. It will be noticed that 65 per cent of this heating oil is "fuel oil"—most of this being "heavy fuel oil"—and that this

---

Fig. 5. The sources, production and distribution of the main forms of fuel and energy in Great Britain in 1974.

Part I The sources of supply of oil and its distribution as various products, of which about half are used for heating purposes and this part is passed on to Part II.

Part II This shows how coal, natural gas and oil along with nuclear energy and water power are used to supply the nation's needs for solid, liquid and gaseous fuels and for electricity. The distribution of this energy among various classes of user is shown and particular prominence is given to the requirements of the steel industry.

Information is based on the Digest of United Kingdom Energy Statistics. In some places estimates have had to be made.

In converting all quantities to coal-ton equivalents the following factors have been used:

$$1 \text{ ton coal} \equiv 1 \text{ ton coke}$$
$$\equiv 0.59 \text{ ton oil}$$
$$\equiv 255 \text{ therms}$$
$$\equiv 7400 \text{ kWh}$$

part can be used only by industrial users with suitable equipment. Electricity and steel between them must use most of these grades. Other industries use the lighter grades such as gas oil while the domestic market and commercial users use the kerosene grades. It is highly desirable that the market is matched to the supply and sudden big changes in either the demand or the supply proportions are not easily accommodated.

Most of the gas distributed now comes from natural gas fields at the southern end of the North Sea. The imported $1 \cdot 1$ mtce comes from Algeria as liquefied natural gas (lng) under contracts entered into before the North Sea gas became available. Some oil and refinery gases and some coke oven gases are also processed and distributed by the gas industry. Town's gas distribution is being phased out and is now down to about 10 per cent of the total in 1975. It is interesting that while in 1960 town's gas was based on coal gas made in retorts a change was made during the 1960s to the production of a similar fuel from the "naphtha" cut from oil refineries. This was changed a second time in a decade when the North Sea natural gas came on stream. A nation-wide gas grid had to be built to distribute this new fuel and appliances converted for its use at all levels. The exploitation of natural gas resources has been one of the most remarkable industrial phenomena in the past 30 years. It started in the U.S.A. where indigenous supplies are now clearly beginning to fail and where the main hope of keeping their vast costly network of pipelines full is to purchase from Canada, but they are also looking to seaborne importations of lng possibly even from Russia. Britain pioneered the transportation of lng by tanker and it is now carried vast distances by sea, much of it to Japan. Pipeline distribution serves Europe from sources in South France and North Holland and there are pipelines also in Russia from Iran and adjacent Azerbaijan. The reserves of this fuel are very limited. In most areas they are not likely to last beyond 20–40 years. While they are available, however, it is important that they should not be allowed to go to waste as so much of it has done in the past in the oil fields.

The contribution by nuclear energy to Britain's total energy requirements has risen since 1960 from 0·23 to 3·1 mtce of electricity produced. This represents about 12 million tons of coal saved. This figure is rising rather slowly and spasmodically. The technology involved is very difficult. Stations built are very large and tend to be built in groups all of a kind followed by a pause while the next "generation" is developed. Each generation of stations is hopefully more efficient than the previous one and it is always a difficult political decision whether to build or wait in the expectation that the later model will always produce cheaper electricity. Different countries have proceeded at different rates and with different degrees of success. It does seem probable, however, that within the next thirty or forty years most of our electricity will be made from nuclear energy and that much more of our distributed energy will be electrical than is the case today. It has even been suggested that distributed gas will be hydrogen, generated by electrolysis and piped in high pressure grids. Only cheap nuclear power would make this possible.

It is of interest to look at some of the details of the changes in the consumption patterns. The steel industry used 22·0 mtce of energy in 1974. This is about 8 million tons less than in 1960. Steel production decreased in the same period from 24·3 to 22·4 million tons. Coke used fell from 14·5 to 9·0 million tons while the iron produced decreased from 15·8 to 13·9 million tons but there has been some substitution here particularly by oil. Nevertheless there has been a significant improvement in energy economy in that industry—on top of the substantial improvements already enjoyed since 1945. By comparison homes used in 1974 58 mtce of energy and other space heating accounts for a further 18 mtce—76 mtce in all for relatively domestic purposes, or over 25 per cent of the total used. This is an increase from 47·6 mtce in 1960 which was then 21 per cent of the total. It should be appreciated that these figures include 12·5 mtce of electricity in 1974 which needed about 40 million tons of coal to produce, so that the total proportion of energy going to provide personal comfort is even higher than indicated above. Counted this way homes alone used 22 per cent of the total internal consumption. Clearly there has been a significant rise in the level of comfort being enjoyed by the population, especially considering the improvements in the insulation of buildings which has been effected in recent years. The use of central heating has, of course, become common in small homes whereas even only fifteen years ago it was quite rare in Britain.

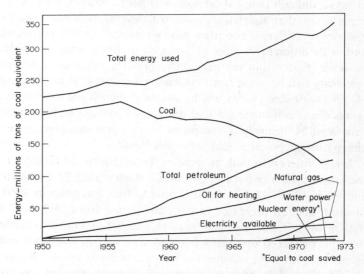

FIG. 6. Trends in Great Britain in the consumption of energy in various forms between 1950 and 1973.

Figure 6 shows some of the trends in fuel utilization over the past years. There is an overall increase in the total energy being used but this has been lower than the general rise in living standards because of the improved efficiency of utilization. New plant of all kinds from domestic fireplaces to power stations have become better designed and modernization of equipment at any time in recent years would almost always lead to lower energy costs per unit of production. As time goes on, however, the law of diminishing returns must apply. Improvements still possible become smaller and more costly to effect. There is, then, still a demand for more and more energy tempered somewhat by a continuing improvement in the efficiency of utilization. It is unlikely that the savings will ever be great enough to meet the additional demand without the need to use extra fuel.

The rate at which efficiency increases cannot be measured outside a few restricted industries. The improved efficiency of the power stations has already been mentioned (page 85) and it is such that if the efficiency of 1939, say, had not been improved upon, our present production of electricity would be costing us not 100 mcte but 150 mcte today. Similarly the improved thermal efficiency of blast furnaces since 1950 is saving us about 8 million tons of coal annually. In each case there is a big saving in the cost of capital equipment too, fewer power stations, blast furnaces and coke ovens being required—although those we do build are not themselves cheap, being increasingly sophisticated in their design. In steelmaking the switch from the open hearth to the basic oxygen process saves a large amount of oil annually. Further improvements are no doubt possible but in electricity generation, the efficiency having risen from about 15 per cent to 38 per cent in the best plant, it can never be increased beyond 45 per cent which is the theoretical limit unless means can be found for converting chemical to electrical energy directly on a large scale as is possible using fuel cells on a restricted scale.

It is difficult for an industrial nation to compete successfully with others if it has an energy policy which leads to higher fuel costs than those of its competitors. For this reason in the 1950s and 1960s increasing quantities of imported oil had to be used in Britain rather than indigenous coal. In the mid-1970s the price of oil makes it less attractive than it has been so that the pattern of use may alter again but these changes are fairly slow processes and the outcome will not be apparent until the availability and cost of North Sea oil is established. The cost of fuel, it should be appreciated, includes the cost of handling within the user's premises including the cost of ash disposal. These considerations set coal at considerable disadvantage against oil and to a lesser extent oil at a disadvantage against gas.

One outstanding feature of Fig. 6 is the rapid development of the gas industry since the discovery of natural gas in the North Sea. The old coal gas industry had been in difficulty for many years and was being up-dated as an oil gas industry when natural gas was found. At first natural gas was reformed to coal gas quality for domestic use or directed on short term contracts to selected large industrial users pending the construction of a new transmission grid of pipelines and the initiation of a programme of appliance conversion. At present gas is the cheapest fuel available in Britain but under Government control costs of different forms of energy are not likely to be allowed to remain significantly different for very long especially for domestic users.

The next big development will be the landing of oil from the North Sea oil fields. Large deliveries are expected by 1977 but the technology is barely developed and costs are likely to be rather high. This oil is expected to be rather light—providing gasoline and kerosene grades in high proportions, and consequently less of the heavy fuel oil grades. It will therefore be low in sulphur and of relatively high value. Unless the consumption pattern changes it will be necessary to exchange some of this oil on the market for heavy grades for use in industry.

The gradual disappearance of coke from the domestic market has been determined mainly by the supply situation. Once the cheapest domestic fuel, coke is now scarce because the gas works have ceased to produce it, and it has become very costly relative to alternatives. Other smokeless fuels are also expensive but are better suited to most domestic purposes. Industrially, coke is still required in cupolas and blast furnaces but its attractiveness elsewhere is diminishing. Most of the coke available for domestic use is hard metallurgical coke rejected from iron works because of size. It is not well suited to household purposes.

The use of electricity has been increasing enormously despite its high price. Metallurgical melting by electricity used to be used only for special purposes such as the production of high alloy steels in small arc furnaces. Today secondary steelmaking from cold scrap is commonly carried out in arc furnaces of 300 tons capacity. In these, melting rates are high and contamination is minimal. Energy costs per ton are high but this is offset by increased production rate which reduces standing charges per ton of steel made. In general electricity is highly priced per unit of energy but can be used so much more efficiently that the price can be absorbed among other costs. The inefficiency has already occurred in the power station. Electricity is flexible in use and continuity of supply is fairly well assured though the small chance of interruption sometimes makes the installation of costly standby equipment necessary. Electricity is of course used for light and power as well as for heating and for these purposes its high cost is readily acceptable. Its supply involves the community in vast capital expenditure to keep production capacity level with increasing demand and there is a perpetual debate about the best system, based on coal, oil or uranium which will provide electricity at the lowest cost during the not inconsiderable life of the next station to be constructed. The decisions are made essentially by the politicians who have to allocate the money. Only historians will be able to find out with what wisdom these decisions are being made today.

Reserves

World-wide reserves of coal are immense and its distribution is widespread. At current rate of use there is clearly sufficient for several hundred years and Britain has an adequate share of these reserves though unfortunately a high proportion of known seams would be costly to mine. Oil reserves are much smaller and less uniformly distributed. The current opinion is that oil from wells will not be freely available beyond about A.D. 2010. It will not by then be unobtainable but it will by then have become scarce and very costly. By that time the oil-sand and shale deposits which are very extensive will be producing oil but production costs will be high and the extent of this development will depend upon the supply, demand and cost situation at that time. By the middle of the next century even these reserves may be running low. Estimates like these depend on two imponderables, however, the rate of discovery and the rate of use. To meet the current increasing rate of demand would require the annual discovery of an oil rich area comparable to the North Sea find of the 1970s. Most new fields are likely to be found in off-shore continental shelf sites where recovery of the oil cannot but be costly. The time is clearly not far off when the use of oil as a fuel will become restricted, probably at first by financial considerations, later by national policies, until eventually it will be reserved for use as a source of carbon, as a reducing agent, as a lubricant and possibly for petrochemicals.

Obviously the energy will have to come from another source. At present the only alternative practicable is uranium. The known reserves of this element in deposits which can be mined cheaply are quite limited and may not be much greater in energy terms than the oil reserves if used in conventional power stations but would extend to provide about as much energy as the known coal reserves if used in breeder type reactors. Beyond that there are uranium ores which would be more costly to work and thorium deposits. There is also uranium in sea water and it may be that the sea will one day be the source of many of man's mineral requirements. Current technology could extract uranium from sea water only at about five times too high a cost. The gap narrows. There remains also the hope that fusion may yet be controlled, in which case the necessary lithium and deuterium would be available in sufficient quantities for thousands of years. The cost would scarcely matter. At a more mundane level there are schemes under discussion for the exploitation of solar energy, geo-thermal energy, tidal energy and even wave energy—all of these quite inexhaustible and each capable of making very significant contributions in the right locations. Already the French have succeeded in harnessing the tide both in Mauritius and in France and it has been estimated that the tidal flow in the Severn could be made to supply about a fifth of Britain's (current) energy needs. Geo-thermal energy is now being tapped only in volcanically active areas but deep boreholes might be able to reach down to limitless supplies of heat almost anywhere. Current indications are, however, that the cost would be extremely high.

Perhaps the most astonishing aspect of the current energy availability crisis which the world seems now to be facing is the rate at which man's capacity for using—or wasting—energy has been increasing in the past few decades. Even in "fully developed" industrial countries the *per capita* demand, already very high is still rising at about 5 per cent per annum while in "developing" countries the rate of increase of energy consumption is even greater as populations increase, industrialization advances and personal aspirations grow. There is presumably a limit to the rate at which energy can be usefully consumed but there is no sign today that that limit is being approached.

PART TWO
Furnaces

10

The Evolution of Heat

THE amount of heat available by the combustion of a fuel depends on its calorific value. The amount of heat actually obtained is less easily assessed. It may be said to depend on the effective calorific value under particular combustion conditions and must take into account the actual combustion reactions and their degree of completeness. Particularly at very high temperatures, both H_2O and CO_2 tend to dissociate so that any assumption that oxidation of H_2 and C is complete will be invalid, and any empirical determination of c.v. in which these are oxidized completely will give a figure which is higher than that realized in practice.

When carbon and hydrogen are both present in the fuel and when the burned gases contain excess oxygen (as they usually do) the equilibrium composition of these gases will depend on the equilibrium constants of the reactions:

$$2CO_2 = 2CO + O_2: \qquad \Delta G_T^\circ = 135,100 - 41\cdot5T \qquad (10.1)$$
$$\text{cal}$$

and

$$2H_2O = 2H_2 + O_2: \qquad \Delta G_T^\circ = 118,000 - 26\cdot75T \qquad (10.2)$$
$$\text{cal}$$

for which the equilibrium constants are:

$$K_{CO_2} = \frac{p_{CO}^2}{p_{CO_2}^2} \cdot p_{O_2} \qquad (10.3) \qquad \text{and} \qquad K_{H_2O} = \frac{p_{H_2}^2}{p_{H_2O}^2} \cdot p_{O_2} \qquad (10.4)$$

these being related to the standard free energies by the van't Hoff equation $\Delta G_T^\circ = - RT \ln K_T$

Obviously the p_{O_2} value is the same in each case and is to a first approximation the value of the partial pressure of the excess of oxygen in the combustion gases calculated by assuming complete oxidation. If the oxygen partial pressure and temperature are

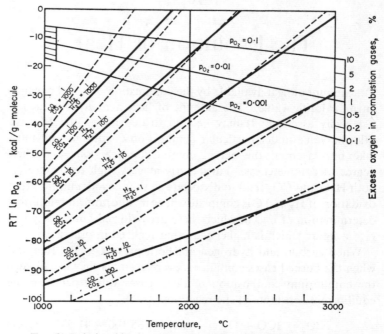

FIG. 7. Relationships between temperature, oxygen partial pressure and the equilibrium ratios $CO:CO_2$ and $H_2:H_2O$ for use in determining the degree of dissociation of CO_2 and H_2O in flames.

known the H_2/H_2O and CO/CO_2 ratios are fixed. The relevant relationships are summarized in Fig. 7 in which the standard free energies of formation of mixtures of CO and CO_2 and of H_2 and H_2O and of expansion of O_2 are plotted against temperature. The intersection of the temperature ordinate and the appropriate p_{O_2} line determines the relevant values of the gas ratios. If the value of p_{O_2} first chosen is only approximate, it can be amended to allow for oxygen produced by the dissociation indicated and a second analysis determined. For example, if CH_4 is to be burned with an excess of 10 per cent of oxygen over the stoichiometric requirement, to produce a temperature 2100°C, the products will have a CO/CO_2 ratio of about $1/10$ and an H_2/H_2O ratio of about $1/100$. The oxygen actually combined would be about 3 per cent less than the stoichiometric requirement so the excess of oxygen would be 13 per cent over the actual requirement. Associated nitrogen (in air) would similarly be in greater excess than intended. This could have an important effect on flame temperature. Up to about 1700°C the effects of this dissociation can probably be neglected for most purposes, but beyond 2000°C they are very important and at 3000°C the further effects of dissociation of hydrogen and oxygen to monatomic gases and of water vapour to monatomic hydrogen and hydroxyl molecules might have to be taken into account. Flame temperatures of this magnitude are now possible using oxygen instead of air. Note that the flame temperature would increase (up to a limit) as the excess oxygen or p_{O_2} value increased because of the effect this would have on the completeness of the oxidation to CO_2 and H_2O.

Flame Temperature

The temperature attained by the combustion of a fuel depends not only on its calorific value but also on the burning technique used. Coal normally burns in an open fire to give a temperature of the order of 1000°C but if the oxygen supply to it is restricted it may smoulder but continue to oxidize at about 200°C. Under more favourable conditions than normal, 2000°C might be achieved.

The factors affecting the temperature attained in the combustion of a fuel are:

1. Its effective calorific value which must depend on the temperature actually attained as this affects the equilibrium of the various reactions.

2. The amount of diluent—usually nitrogen—admitted with the oxygen. Using the normal proportions of oxygen and nitrogen in air necessitates imparting about three quarters of the heat to the nitrogen. If the air is enriched with oxygen or if it is replaced by so-called "tonnage" oxygen (99 per cent) the amount of heat released will be unaltered but temperatures attained will be much higher. Any excess of air over the stoichiometric requirement acts as a diluent, of course, and moisture in it may dissociate endothermically. Ash affects temperature in a like manner.

3. The temperature of the fuel and air prior to combustion. The sensible heat in the reactants should be added to the calorific value in assessing the total heat available in the combustion products.

4. The rate at which the reactions take place. Ideally if the combustion occurs instantaneously, all the energy is released as heat and is taken up by the products in no time and therefore without any heat loss having taken place by radiation or otherwise so that the heat content of the products is a maximum and therefore their temperature is a maximum. In practice, of course, this is not possible but the closer the ideal is approached the higher is the temperature that is attained. In practice heat is being lost *during* the reaction so that the build-up of heat content in the reaction products falls short of the maximum, and the longer the time during which the evolution of heat proceeds the greater are these simultaneous heat losses incurred. While the energy is not necessarily lost to the system, the maximum temperature is not being attained. The rate of the reaction can be increased by suitably designing the burner or, where appropriate, by increasing the reactivity of the fuel. Burner design should be directed toward more rapid mixing of the

fuel and oxygen. This may be done in various ways appropriate to different fuels, and will be dealt with in detail later.

It will be observed that of these factors, (1), (2) and (3) are quantities which can be measured and controlled to desired values. After they are fixed, however, the temperature attained is still very dependent on the rate of the reaction and hence on the burner design.

It is possible to calculate flame temperatures. An *ideal* flame temperature can be determined by assuming (1) complete oxidation by the exact stoichiometric addition of air (or of oxygen with stated nitrogen accompanying it), (2) perfect mixing, and (3) instantaneous combustion so that no heat losses occur during burning. Then the total heat content of the reactants plus inerts (N_2, ash) above room temperature is calculated on the basis of unit mass or volume and to this is added the heat of the reaction. This sum is taken to be the heat content of the products. The volume (or mass) of the products is then calculated and hence their temperature, using an appropriate value of specific heat (or of integrated "total sensible heat at $T^\circ C$" values which are tabulated in reference books[4]).

Example.   Calculate the ideal flame temperature for the combustion of methane by air preheated to 600°C.

Consider the combustion of 1 mole of methane. The heat of oxidation is given by

$$CH_4 + 2O_2 = CO_2 + 2H_2O; \Delta H = -191{,}800 \text{ cal}$$

Oxygen required = 2 g molecular volumes = 44·8 l.
Air required   = 44·8 × 100/21   = 213·5 l.
Sensible heat in air at 580° over ambient = 213·5 × 580 ×
                                                              specific heat of air
                                                              (= 0·324 cal/l./°C)
                                                            = 40,000 cal
Total heat in products of combustion   = 231,800 cal

Products consist of 22·4 l. CO_2, 44·8 l. H_2O and 169 l. N_2 = 236·2 l. in all. The specific heats of CO_2, H_2O, and N_2 are respectively 0·58, 0·46, and 0·36 cal/l./°C. These are mean values between 0° and 2000°C, the expected approximate value of the flame temperature.

The weighted mean value of the mean specific heat of the combustion gases is therefore

$$(22\cdot4/236) \times 0\cdot58 + (44\cdot8/236) \times 0\cdot46 + (169/236) \times 0\cdot36$$
$$= 0\cdot40 \text{ cal/l./°C}$$

The temperature rise is therefore $\dfrac{231,800}{236 \times 0\cdot40} = 2450°$

The flame temperature is therefore 2470°C (adding the ambient 20°).

The similar calculation for a commercial fuel would use the calorific value instead of the heat of oxidation and allowances would have to be made for the effects of other components such as sulphur compounds, moisture and other hydrocarbons on the product volume and heat capacity.

This ideal flame temperature is higher than can be achieved for several reasons—particularly that assumptions (2) and (3) can never be valid. Assumption (1) could be applied in practice but would usually result in an undesirably slow completion of the reaction, and a calculated excess of air is always used, the amount (up to about 50 per cent) depending on the kind of the fuel.

A modified value of the flame temperature could readily be calculated allowing for any required excess of air—both O_2 and N_2 being carried as diluents.

Example. If in the previous example the product were to carry an excess of 2 per cent by volume of unburned oxygen, what would the adiabatic flame temperature then be?

The sensible heat in the products would be increased by the heat introduced by 4·72 l. of oxygen and the associated 17·8 l. of nitrogen at 600°C i.e. by 4230 cal to 236,030 cal.

The volume of the products would increase by 22·5 l. to 258·5 l. and the mean specific heat of this gas would be almost unchanged so that the temperature rise would be

$$\frac{236,030}{258\cdot5 \times 0\cdot40} = 2290°$$

The adiabatic flame temperature is therefore 2310°C.

If an estimate can be made of the heat lost from the combustion zone during combustion a non-adiabatic temperature can be calculated which will of course be no better than the estimate made. Such a calculation might be useful when a burner was being designed to help to quantify the advantages of various design alternatives.

Example. If in the previous example, 5 per cent of the total heat available is deemed to be lost during combustion— a small proportion in practice—how would the temperature be affected?

Clearly the temperature rise would be reduced by 5 per cent to 2170° so that the calculated flame temperature would be lowered to 2190°C. This is not, of course an ideal or adiabatic temperature but the result of an attempt to calculate an actual flame temperature.

A further desirable correction would be to make allowance for the modification of the calorific value due to partial dissociation of the reactants—or incomplete oxidation of the fuel—as has been discussed above. This would be done as follows:

Calculate the ideal flame temperature as above, allowing for excess air and use this as a first approximation. Calculate the equilibrium proportions of reactants and products at this ideal temperature using Fig. 7. Estimate a better value of calorific value in the light of this information. Recalculate the flame temperature to a second approximation. Repeat to a third and perhaps a fourth or fifth approximation until successive values are nearly the same.

Such a calculation would obviously be long and tedious. Several graphical aids to its completion have been published, one of which is to be found in reference (4) of the bibliography.

The direct calculation can readily be handled by a simple computer programme by means of which the effects of a number of independent variables can be examined separately or together. Reference to Fig. 7 indicates that when there is a 2 per cent excess of unburned oxygen and a temperature of about 2200°C the equilibrium CO/CO_2 ratio will be about 0·25 and the H_2/H_2O ratio about 0·03. This clearly indicates that combustion, even in this case will not be complete, and that the temperatures calculated above will be for this reason alone, rather high. When full allowance is made for this effect it can be shown that the flame temperature with 2 per cent excess oxygen and 5 per cent heat loss would be only 2154°C. In these circumstances the CO/CO_2 ratio would be 0·21 and the H_2/H_2O ratio 0·03 while the net heat in the products of (partial) combustion would be only 205,966 cal instead of the 224,307 cal when the oxidation is complete.

Another example of incomplete oxidation is the combustion of coke in the blast furnace where the excess of carbon present restricts oxidation to the formation of CO only in the tuyere zone. Consequently only about a third of the calorific value of the fuel is released. To obtain a reasonably high flame temperature it is necessary to preheat the air blast, usually to about 1000°C. The coke also reaches the combustion zone hot, at about 1400°C, yet the temperature attained is not above about 2000°C. A further lowering of this temperature occurs if moisture is present in the air blast due to its endothermic dissociation. The effect of any such reaction on the flame temperature can readily be estimated by making simple amendments to the basic flame temperature calculation.

The results of such calculations would be very useful when, for example, one was examining a new fuel or developing a new burner and wished to assess one's success in approaching the theoretical value of flame temperature. A final development of the calculation of flame temperature would be to assess the effects of loss of heat from the body of combustion gas while it remained in the furnace and hence its fall in temperature. It would be necessary to divide the furnace into several arbitrary zones and to consider the heat developed and dissipated in each zone. In the first zone the temperature would obviously be raised by the development of a great deal of heat as combustion commenced. The final temperature as the gases left that zone would be calculated along with the gas analysis at that point, allowing for heat losses based on the average temperature of the gases in Zone 1 and the time taken to traverse it. In Zone 2 combustion reaction might well continue and the flame temperature at the end of Zone 2 would depend on the heat in the gases entering it plus the heat of further reactions minus heat lost during the traverse of Zone 2. This could be continued through the furnace. In later zones there would be no heat of further reaction and heat loss to the stock would depend on whether or not cold stock was being encountered. This is in effect a heat balance by stages—a very valuable analysis of a process if sufficiently accurate information can be assembled to make it a success.

Available Heat

The advantage of achieving high flame temperatures becomes clear if we consider the concept of available heat. This is simply an expression of the 2nd law of thermodynamics. If the temperature of a body of combustion gas is T_2 and its mean specific heat between 0 and $T_2°C$ is C_p per cubic metre (at STP) then the sensible heat content of the gas is $C_p.T_2$ per m³. If the gas is cooled to T_1, by virtue of giving up some of its heat to raise the temperature of an element of burden from $(T_1 - \delta T)°$ to $T_1°$ its heat content becomes $C_p.T_1$ (assuming C_p not appreciably altered) and the heat available in the original gas for doing useful heating at T_1 was obviously $C_p.(T_2 - T_1)$ per m³.

It should be equally obvious that if by different combustion techniques the same fuel had been burned to produce gas at T_3 ($>T_2$) then the available heat at T_1 would have been $C_p(T_3 - T_1)$ per m³. Particularly when T_1 is high it is very important that the combustion gases should be as hot as possible so that as high a proportion as possible of the total calories shall be available at the working temperature.

For example, if the burden in a furnace or kiln is to be held at 1600°C a flame at 1700°C would be useful only in so far as it could yield up heat as it was itself cooled down to 1600°C, i.e. only about one-seventeenth of its heat content is available at 1600°C. A flame at 1800°C would, however, be able to yield up about one-ninth of its heat above ambient to maintain the kiln at 1600°C while a flame at 2000°C would have one-fifth of its heat available at 1600°C so that only about one-third of the fuel would be required to have the same effect as with the 1700°C flame. The importance of flame temperature, especially where high temperatures are to be developed, is obvious.

Graphical representation of the fraction of total heat available at $T°$ can readily be made from specific heat data and affords a convenient means of comparing either fuels or combustion conditions (Fig. 8). These graphs can also be extended to include the "demand" of the burden in cases of counterflow heating in such a way as to show in what areas the gas–burden temperature differences will be high or low and the heat transfer rates correspondingly high or low.

The "Virtue" of Energy

Thring[15] develops a further concept which he calls the "virtue" of energy. Briefly the virtue V_1 of a quantity of energy Q all at the same temperature $T_1°$K over its surroundings at $T_0°$K is given by

$$V_1 = Q\left(1 - \frac{T_0}{T_1}\right) \qquad (10.5)$$

this being the maximum part of the energy which is available for doing work in an ideal heat engine operating reversibly between the temperatures T_1 and T_0 on the thermodynamic scale. This follows from the Carnot cycle treatment of the efficiency of a reversible heat engine, to be found in any thermodynamics text-

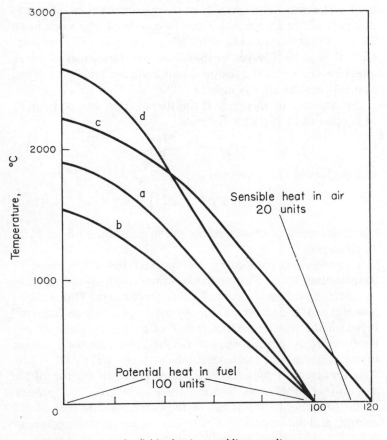

FIG. 8. Available heat diagrams for products of combustion of coal under several conditions; (a) with theoretical cold air, (b) with excess cold air, (c) with theoretical preheated air, the preheat supplying 20 per cent additional heat, and (d) with theoretical cold, oxygen enriched air. The temperature at which available heat is zero is the flame temperature. The exact form of the curves depends on the specific heat curves of the combustion gases.

book. The virtue of a quantity of energy is defined by Thring as that part of the energy which can be converted into work in an ideally perfect system. The virtue of the energy in a body of combustion gases at T_1° over surroundings, say furnace stock at T_0° might be said to be the thermodynamically available part of the available heat as already defined.

If a quantity of energy Q is transferred from one body at T_1 to another at T_2 its virtue becomes

$$V_2 = Q\left(1 - \frac{T_0}{T_2}\right) \tag{10.6}$$

and the loss of virtue suffered is

$$\delta V = Q\left(\frac{T_0}{T_2} - \frac{T_0}{T_1}\right) \tag{10.7}$$

which increases, of course, as the fall in temperature from T_1 to T_2 increases.

In the general case energy occurs over the whole range of temperatures up to the measured value and it is necessary to summate the values of virtue over this whole range. Thring shows how this can be done graphically by means of a "virtue diagram" in which the fraction or percentage of energy available at any temperature is plotted against the negative reciprocal of the temperature in the thermodynamic scale $(-1/T^\circ K \ (T > T_0))$. This arrangement puts the high temperatures to the top of the diagram as in Fig. 9. Between these axes areas are proportional to values of the virtue of the corresponding part of the energy, and differences in virtue due to varying conditions or caused by reactions or transferences of heat can readily be observed qualitatively or computed by planimetry or by counting squares.

This can readily be appreciated by considering the area of a narrow vertical strip under the curve, δ_q wide and reaching from T_0 to T_1. The area is

$$\delta q \times \left[\left(-\frac{1}{T_1}\right) - \left(-\frac{1}{T_0}\right)\right],$$

that is the difference between the area of a strip reaching from $-\infty$, corresponding to $T = 0°K$, up to the T_1 level and that of a similar strip reaching only to the T_0 level. This is clearly equal to $\delta q/T_0(1 - T_0/T_1)$ or $(1/T_0) \times$ Virtue. By graphical integration

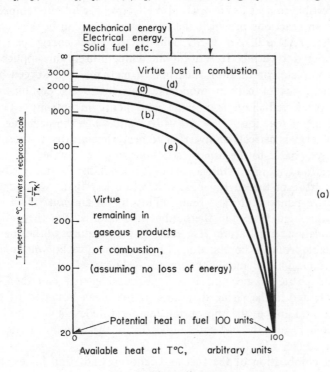

FIG. 9. Virtue diagrams.
Virtue diagrams for conditions (a) (b) and (d) of Fig. 8. Curve (e) shows loss of virtue on dilution of condition (a) with an equal volume of cold air.

the area under the curve is proportional to the virtue of the energy so represented.

Because mechanical and electrical energy are perfectly convertible to heat in accordance with the first law of thermodynamics the virtue of these forms of energy is equal to the amount of energy—i.e. $Q(1 - T_0/T_E) = Q$ so that the effective temperature of the energy $T_E = \infty$ in each case. Chemical energy is slightly different. In so far as Gibbs's free energy change ΔG is perfectly convertible to heat its energy is also effectively at infinite temperature. In reactions in which the entropy change can be neglected so that $T\Delta S \simeq 0$, $\Delta H \simeq \Delta G$, and the chemical energy can be considered to be effectively at infinite temperature. This applies to the complete combustion of carbon or methane with oxygen but in the cases of carbon monoxide and hydrogen, combustion involves a reduction in volume, and hence in entropy. Then the heat of reaction $\Delta H = \Delta G + T\Delta S$ and while the part due to free energy is perfectly convertible to heat, the part due to thermal energy is liable to the usual restrictions imposed by the 2nd Law of thermodynamics. Only a part of it is available for doing work. The virtue of the energy ΔH is therefore less than ΔH and the effective temperature less than infinity by an amount which can be calculated from the thermodynamic data at the theoretical combustion temperature. The effective temperatures of the energy available from the combustion of hydrogen and carbon monoxide are about 5400°C and 3200°C respectively.

The virtue in mechanical and chemical energy can then be represented in the virtue diagrams by the areas bounded by the horizontal lines at the appropriate temperature levels, the abscissa through T_0 and the ordinates indicative of the amount of energy involved (see Fig. 10).

On combustion of the fuel and ignoring radiation losses, the chemical energy is transformed to sensible heat in a body of gas at the theoretical flame temperature. Its virtue is then represented by a hyperbola such as curve (a) in Fig. 9, the exact form of which depends on the specific heat curve for the gas or gaseous mixture. It is obvious from the form of this curve that a large loss of virtue has occurred. The use of excess cold air for combustion reduces the virtue still further (curves (b) and (e)) while oxygen enrichment reduces the loss of virtue (curve (d)). In Fig.

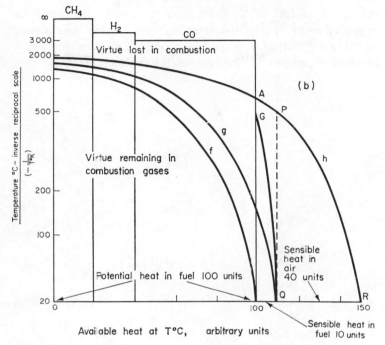

FIG. 10. Virtue diagrams.

Virtue diagrams for combustion of a gaseous fuel in which 20 per cent of the heat comes from CH₄, 20 per cent from H₂ and 60 per cent from CO; (f) without preheat of gas or air, (g) with 10 per cent extra heat available as sensible heat in the gas and (h) with a further 40 per cent as preheat in the air. The temperature of the preheated gas is read at G but A indicates a weighted mean temperature of gas and air. The virtue represented by PQR is recycled continuously in the preheating system.

10 curve (h) shows the effect on virtue of using preheated air, when compared with curves (f) and (g), drawn for cold air.

It is obviously desirable that the virtue, or quality or potential of energy should be obtained and maintained as high as possible during its useful "life". Once heat is degraded it cannot be upgraded and it is very important that even "waste" heat should be taken off with as high quality as possible if there is any way in which it can be re-cycled or re-used (see also p. 174).

11

The Combustion of Fuel

Solid Fuel Beds

Coke

THE commonest solid fuel is carbon in the form of coal or coke and the combustion of coke in beds has been most fully described by Mott and others in the 1930's.

The process appears to occur in two stages although these may overlap to a considerable extent. Observation of gas analyses at various distances from the supporting grate shows that the distribution of molecular species in the gas phase when a hard, unreactive coke is burned is like that indicated in Fig. 11(a).

In the first stage

$$C + O_2 = CO_2; \qquad \Delta H = -94,200 \text{ cal}$$
$$= -394 \cdot 3 \text{ kJ} \qquad (11.1)$$

appears to be the predominant reaction and only small amounts of CO are detectable.

In the second stage, when all the oxygen has been consumed

$$C + CO_2 = 2CO; \qquad \Delta H = 40,200 \text{ cal}$$
$$= 168 \cdot 3 \text{ kJ} \qquad (11.2)$$

until all the CO_2 is consumed.

The space in which the first stage occurs is about 3 particle diameters from the supporting grate (if the fuel is uniformly sized). The space in which the second stage occurs will be at least as much and possibly very much more, to the point that CO_2 may

121

escape unconverted from the top of the bed. This condition is most likely if the reactivity of the coke is low.

In so far as reaction (11.1) is exothermic and reaction (11.2) endothermic, the maximum evolution of heat has occurred at the end of stage (1) and in the operation of a furnace using this type of burner, stage (2) should be inhibited as far as possible by using fuel with a low effective reactivity and a short fuel bed.

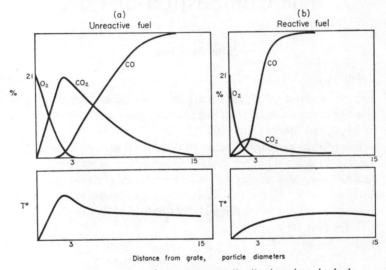

FIG. 11. Gas analyses and temperature distributions in coke beds. (a) very unreactive fuel. (b) very reactive fuel.

If a very reactive fuel is burned, fuel bed analysis gives results shown in Fig. 11(b) where the solution of carbon by CO_2 has commenced almost as soon as CO_2 has formed. The maximum temperature is not so high as with unreactive fuel, nor so sharply defined.

The effective reactivity of a fuel bed may be increased by:

1. Increasing the actual reactivity by modifications in the coke production process, for example by using a lower coking temperature.

2. Raising the pre-heat in the coke and/or the air.
3. Using smaller sized material, i.e. having larger specific surface. (This is complicated by the fact that the permeability of the fuel bed then falls and hence the air–gas mass flow rate is reduced. If this is restored by increasing the pressure differential across the bed the linear flow rate is increased which would also speed up the reaction rate by thinning the gas film on the solid surface.)

Re-examining the reactions, we find that according to Strickland-Constable* who investigated the reaction of low pressure oxygen on incandescent carbon filaments, CO is probably the primary product of the reaction with carbon, and it is also the final equilibrium product in the presence of excess carbon. ($CO/CO_2 = 10^4$ at 1500°C.) It seems most probable that a thin film or envelope of CO encloses the coke particle at ordinary pressures, and is substantially in equilibrium with it.

Reactions:

$$C + O_2 = CO_2 \tag{11.1}$$

and

$$C + CO_2 = 2CO \tag{11.2}$$

can then occur only by the O_2 and the CO_2 molecules penetrating this film either by impact or diffusion. The film can be presumed thinner at very high temperatures and when wind speed past the fuel is high, but it is probably impossible to destroy it. It is possible that reaction (11.1) does not take place under these conditions to an appreciable extent. The CO_2 is produced by the reaction:

$$2CO + O_2 = 2CO_2; \quad \Delta H = -134,400 \text{ cal} \tag{11.3}$$
$$= -562 \cdot 6 \text{ kJ}$$

which occurs freely in the void space, its rate depending on temperature, and on the partial pressures of the reactants. In the presence of excess O_2 the equilibrium is reached when almost all

* STRICKLAND-CONSTABLE, R. F., *Trans. Farad. Soc.* **44**, p. 33 (1944).

the carbon is as CO_2. Obviously, simultaneous equilibrium with carbon and oxygen is impossible.

The rate of formation of CO is therefore dependent on the area and effective reactivity of the coke surface, and is likely to be slow due to the thin film of CO over the surface while the oxidation of CO to CO_2 in the voids is probably very fast as long as oxygen is available. Therefore if the fuel is in an effectively unreactive state, CO_2 will predominate in the gases while oxygen remains, (Fig. 11(a)) but if the fuel surface is very reactive or very extensive, reaction (2) may occur fast enough to maintain an appreciable CO content in these gases even immediately above the grate (Fig. 11(b)).

In practice, where the development of a high temperature is the primary objective, large hard unreactive coke (traditionally beehive) is used and the bed is short enough that only a small amount of CO is found in the gas at the top of the fuel bed. If extremely reactive carbon could be effectively burned very rapidly to CO_2, even higher temperatures would be obtained but, so far, such a fuel and suitable burners have not been made. In the other cases, however, the fuel beds are required to produce CO as a secondary fuel (producer gas) or as a reducing agent (blast furnace). In the gas producer small reactive fuel is used. In the blast furnace the higher temperature of the air and the pre-heat in the coke are sufficient to ensure such a high temperature that conversion to CO is complete in a few feet from the tuyere, despite the use of very unreactive fuel.

The small amount of volatile matter in coke must be released in the combustion process and can probably be assumed to enter the gas phase in the voids and react quickly as long as there is an excess of oxygen. When the coke is pre-heated as in the blast furnace the volatile matter may be released so high in the furnace that oxidation is impossible and its potential heat may well be lost, at least to that stage in the process. Reactions involving hydrogen in the volatiles or from moisture in the air are considered in the next section on coal.

As coke is consumed it leaves behind a light friable skeleton of ash which usually crumbles under the weight of fresh fuel moving into the combustion zone. During burning, however, this ash may form quite a thick layer on the surface which must have an appreciable effect in stabilizing the envelope of carbon monoxide around each particle and therefore in lowering the effective reactivity. It is not, however, suggested that a high ash content should be used to achieve this end.

Coal

The combustion of bituminous coal is obviously more complex than that of coke particularly with respect to the volatile matter exuded by most coals as they are heated through the range 200°–400°C—that is before the combustion front arrives. These volatiles contain gases like hydrogen, methane, carbon monoxide and light paraffins and other hydrocarbons which will oxidize rapidly in the gaseous phase wherever there is hot oxygen. There are also heavier tarry materials which may be present as a fine mist and may burn more slowly and provide the flame with its luminosity—or smokiness.

These substances bring hydrogen into the system and, after combustion, water vapour. In the void space the reactions with excess oxygen will tend to form not only CO_2 but H_2O.

$$CH_4 + 2O_2 = CO_2 + 2H_2O; \quad \Delta H = -191,800 \text{ cal} \quad (11.4)$$
$$= -802 \cdot 8 \text{ kJ}$$
$$2H_2 + O_2 = 2H_2O; \quad \Delta H = -115,600 \text{ cal} \quad (11.5)$$
$$= -483 \cdot 9 \text{ kJ}$$

The water vapour will then take part in the reactions at the solid surface,

$$H_2O + C = H_2 + CO; \quad \Delta H = +31,400 \text{ cal} \quad (11.6)$$
$$= +131 \cdot 4 \text{ kJ}$$

along with reaction (11.2). Hydrogen will accompany carbon monoxide in the associated gas film through which the reactants must diffuse. These reactions proceed toward an equilibrium which depends on the temperature being attained, and the amount of excess oxygen. The part played by ash in slowing up the surface reactions must be similar to that in coke combustion—perhaps rather greater because coal ash is more coherent than coke ash.

Coal is more reactive than coke. This is an inherent chemical property. The volatile part obviously reacts very quickly once released and the non-volatile residue is inevitably a low temperature coke which is always more reactive than high temperature coke. Both coal and this coke have a very extensive "internal" surface areas. Even if only a small part of this, near to the visible surface, is available for surface reactions, its contribution to inherent reactivity must be considerable.

Pulverized Fuel

Coal is often used in pulverized form, suspended in an air stream and injected into the furnace along with a stream of hot air—dealt with rather like a gas. Under these conditions it behaves like an extended fuel bed in which the void space is very large even when compared with the surface area, i.e. the ratio specific volume: specific surface is much higher than in an ordinary fuel bed. The void reactions have therefore a much greater chance of occurring than the surface reactions.

The coal particles carbonize, exuding volatiles which form an atmosphere round them which is believed to slow up their reactions with the air, particularly as the relative velocity of the particle and the air stream is practically zero through most of the furnace chamber. Consequently the associated gaseous film on the particle surface is thick and stable and diffusion of reactants across it is slow. In fact, the reactions all occur quite fast, but the individual particles are swept through the furnace in only one or two seconds, and compared with the combustion of gas or oil, pulverized fuel is considered as burning very slowly.

Since volume reactions occur more readily than surface reactions, the fuel burns as if in a very unreactive condition, favouring the formation of carbon dioxide and the development of maximum heat per unit of carbon. The conditions are ideal for developing high flame temperatures provided combustion can be completed very quickly with a small excess of air. This is most likely with low ranking coal (volatile matter > 30 per cent) ground very fine (75 per cent—200 mesh (76μ)). This might need only 20 per cent excess air—more than oil but less than lump coal—and would give a very hot flame. Higher ranking coals and coarser fuel would give longer cooler flames.

During burning the coal particles become as hot as the gas surrounding them and cause the flame to be visible and luminous with correspondingly good heat transfer properties (see p. 170).

The use of pulverized fuel is relatively new and may yet be the subject of much development. It has been adopted by power stations on a large scale. Its main advantage is that it can utilize small fuel and low grade coals, lignites and even peat, but preparation costs must be absorbed when these lean fuels are used.

Metallurgical applications of pulverized fuel are not numerous and its use in the metal manufacturing industries is diminishing probably because of the fact that other fuels are now more readily available, and because sulphur and ash contents can be embarrassing, especially if the ash is fusible. It is sometimes used in reheating furnaces and in similar applications where the ash is not likely to frit. About 85 per cent of the pulverized fuel used in Britain is burned in large power stations and much of the rest in cement kilns.

One type of burner is sketched in Fig. 12. Primary air cannot usually be pre-heated because of explosion dangers, but secondary air is, and oxygen enrichment would seem to offer appreciable advantages.

Coal which is only crushed to about $\frac{1}{8}$ in. (which is much cheaper than pulverized) has been burned in special water cooled combustion chambers in which the fuel and air are forced to swirl round in a very fast tight vortex before issuing axially into the

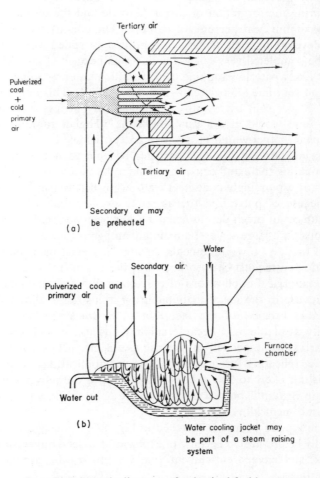

FIG. 12. Schematic diagrams of pulverized fuel burners.
(*a*) showing how grid type intermingles coal and air. (*b*) "cyclone"
combustion chamber for coarser fuel.

furnace proper. The larger particles go to the outside of the vortex where they have longer to burn. They may adhere to the walls on which sticky slag accumulates. Combustion would then be very fast when the relative velocity of the air to the particle suddenly increased. Most of the slag melts and drains to a tap hole in the bottom of the furnace. This type of burner is called a "cyclone" (see Fig. 12(b)).

Oil

Oil may be burned in one of two ways.
1. Vapourized before ignition so that it burns like a gas.
2. Broken into fine droplets which are injected into hot air so that they evaporate while burning.

In the first case the apparatus must be something like a paraffin Primus stove or Tilley lamp and oils must be refined, of high volatility and non-charring. This type is not much used at present on an industrial scale, where the cheaper heavy oils can be used.

In the second case heavy oil (or other liquid fuel) is warmed to a suitably low viscosity and "atomized",
  (a) mechanically by means of a rotating disc or cup or swirler;
  (b) by high pressure ejection from a fine orifice; or
  (c) (and most commonly) by entrainment in a blast of air or steam.

These atomizers differ as to the size of droplet and the shape of the burning zone. (a) gives a wide spray of oil and so a wide flame area but with a very uniform droplet size (50 μ); (b) gives a conical spray; (c) gives a narrow jet and a long pointed flame and produces the most variable droplet size of the three.

Oil (and other heavy liquid fuels) is being used in increasingly large amounts in open hearth steelmaking and in other metallurgical applications. The atomization is almost always by steam in spite of the fact that air atomization would give a better theoretical flame temperature. About 2–8 lb of steam is required per gallon of oil (8 lb). Steam is used because it is cheaper to produce than high pressure air, (or is available almost free), and it also carries some heat which prevents chilling of the oil in the burner. The oil is injected into the hot furnace accompanied by a stream of pre-heated air. The droplets immediately evaporate, and most of the combustion probably occurs as vapour phase reactions. It is not certain whether the droplets evaporate gently or whether they explode. Small particles of char with ash must be formed. These are much smaller than particles of pulverized coal and oxidize readily, but do survive long enough to confer luminosity on the oil flame as is discussed later on page 170. If oil droplets are too big, coke cenospheres may form and resist oxidation within the furnace space.

The shape of the combustion surface is important as it determines the rate at which the oil jet and the air will intermingle and react, and hence the temperature attained. For fast reaction and high temperatures the burner, that is the steam injection *and* the air stream taken together should be designed so that the two streams cut into each other with a high relative velocity and annihilate each other. The shape of the combustion surface must also conform with the shape of the furnace chamber and this factor decides that the long flamed injector type be used in steel furnaces. Ideally, a burner should be variable to give hot short flames or long cool flames as required. In practice this effect can often be attained by controlling the oxygen content of the air, oxygen enrichment giving the hotter, shorter flame.

In some furnaces the oil and air are made to mix and react in a special combustion chamber between the burner and the furnace chamber itself. Fuel and air are forced into a tight vortex in this small compartment. A high degree of turbulence is induced and very little heat is lost from the gases before they flow out, fully burned and at maximum temperature into the furnace proper. One application of that technique is to compact oil fired boilers for land or marine use.

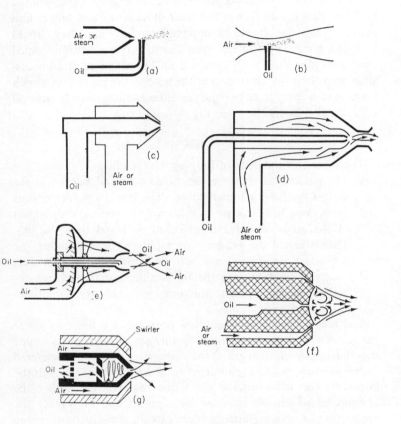

FIG. 13. Schematic diagrams of some oil burners—(*a*) scent spray
type, (*b*) venturi burner, (*c*) high pressure blast burner (narrow
mouth), (*d*) low pressure blast burner (wide mouth), (*e*) mechanical
spreading of oil by rotating cup driven by air impeller; the air
stream then disrupts film into fine drops; cup may be driven
electrically, (*f*) the shape of the nose can be made to induce vorticity,
(*g*) swirlers give oil additional spread on emergence.

Some typical oil burners are sketched in Fig. 13. They should be fairly simple and robust but their dimensions and angles and delivery velocities should be designed carefully and they should be made accurately. In the open hearth furnace a water cooled extension tube directs the aersol from the atomizer at an accessible point into the air stream at the nose of the port. Some such burners also carry in coke oven gas either simultaneously with oil or as an alternative to it (see Fig. 34(c)).

Gas

There are two methods of burning gaseous fuels:

1. The gas and air may be pre-mixed cold and burned at the end of the pre-mixing chamber. The Bunsen burner and its many modifications are of this type. In general only a part of the air needed for combustion is mixed with the gas. This is called the primary air supply. The remainder, the secondary air, is entrained into the flame within the furnace.
2. Gas and air flow into the furnace separately and mix together as combustion proceeds, in what are sometimes called "diffusion flames".

Gas is usually delivered at a low pressure of a few cms water gauge. Air may be drawn into the furnace through ports beside each burner by the draught of the chimney or it may be delivered under pressure beside each burner or at least in part through the burners. The more intense the flame required the higher the pressure at which the air must be supplied and the higher the proportion of primary air. In these cases it is a great advantage to use a gas/air proportioner—a device which ensures that the gas/air ratio is correct whatever the actual flow rate may be. This may be actuated simply by passing the air through a Venturi throat. The reduced pressure in the vena contracta draws the gas into the air stream through a ring of orifices and also determines the position of a piston carrying a sleeve valve which controls the gas flow rate (see Fig. 14(g)). A wide range of flame length can be obtained using such a device if required.

Type (1) is used in small units where it affords the best control over temperature. In the bunsen burner air is entrained by gas issuing from a jet opposite the air valve, into the mixing chamber.

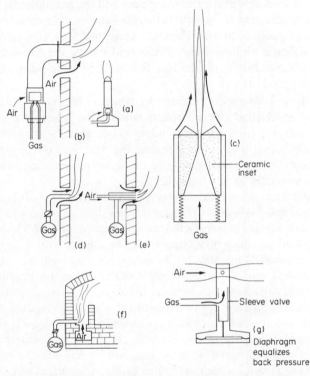

FIG. 14. Schematic diagrams of some gas burners—(*a*) Bunsen burner, (*b*) industrial version of Bunsen burner, (*c*) jet burner, (*d*) simple induced air arrangement, (*e*) as (*d*) but with some primary air supplied, (*f*) induced air supplied through flue may be preheated, (*g*) gas–air proportioner (see text).

The design of the mouth of the burner is important as the linear velocity of the issuing gas–air mixture must exceed by a small amount the flame propagation velocity for the mixture used. If flame propagation velocity is too high the burner will "strike back". For this reason burner mouths may be narrowed to slits so that the issuing gas has a sufficient speed.

A stable flame is attached to the rim of the burner by virtue of the gas velocity at the rim being zero and the paraboloidal form of the combustion front reflects the gas velocity differential across the cross-section of the burner. Attachment to the rim also depends on a chilling effect of the cold rim of the burner as is shown, for example, by the fact that striking back occurs if the rim gets too hot.

In these burners the primary air which is pre-mixed with the gas is insufficient for complete combustion. Secondary air, usually drawn from the surrounding atmosphere, also enters the flame, diffusing inwards to complete the oxidation of the gas. If the theoretical mixture were used the flame propagation velocity would normally be impracticably high.

If the issuing velocity is very high the flame may become detached and blow off. Combustion may then be controlled on a hot brick surface a few inches away. Such a surface would be maintained at a very high temperature in contact with a very hot flame produced by almost instantaneous combustion. In an alternative form of surface combustion the gas–air mixture is passed through a porous ceramic diaphragm. Combustion may take place at the surface or beneath the surface depending upon the gas velocity and the physical parameters involved, such as the thermal conductivity of the ceramic. If the stoichiometric proportion of air is mixed with the gas, combustion may be totally within the ceramic which may glow bright red without any flame being apparent.

Meker burners entrain a high proportion of air near to the stoichiometric proportion. In this case the high flame propagation velocity is accommodated by burning the gas at a deep metal grid which cools the mixture at the point of attachment by conduction and divides the gas stream into a large number of narrow high velocity jets, each burning as a stable flame on the surface of the grid.

Where the air flow is not induced but blown in under pressure, it is possible to approach stoichiometric proportions provided the velocity of the mixture into the combustion zone is adequate. In these cases there must be a properly designed proportioner to control the gas/air ratio.

There is, however, a recent development in this type of burner in which the theoretical proportions are actually used, pre-mixed and passed through the burner with an extremely high velocity, sufficient to avoid a strike-back even though an explosive mixture is being used. The flame is kept "alight" by means of an electric discharge (using a sparking plug) and near theoretical flame temperatures are said to be attainable. Compact standard units can be used for all sorts of purposes, the required temperature being obtained by dilution of the flame gases with air before use, but the greatest technological advantage must be where the very high temperature attainable can be exploited. This type of burner is a recent development with great potentialities. It could not be used with pre-heated air but could be adapted for oxygen and high calorific value fuel as a source of intense heat for metallurgical purposes, and there are reports of "fuel injection lances" or "sonic" burners being used on open hearth furnaces with resulting very high steel production rates.* These burners are, unfortunately, extremely noisy.

Applications of simple Type (1) burners are limited by the size of burner to which stable flames can be attached and by the fact that pre-heated air (or gas) cannot be used because of pre-ignition.

Type (2). Combustion in diffusion flames occurs when gas is allowed to burn at an orifice by reaction with the surrounding air (e.g. the luminous flame of a bunsen burner with the air valve closed) or when gas and air streams are passed into the furnace, in either parallel or impinging streams.

In one class of burners of this type, called jet burners, the gas is injected into the furnace space through a specially shaped slit

* FERRIS, G. A., *J. Met.* **3**, 4, p. 298 (1961).

which produces a turbulent jet of gas which in turn entrains air in a flow pattern determined by the front surface of the burner. These burners may be made of ceramic and they produce characteristic flat flames typified by the bat's wing burner used for glass working. They can be made in large industrial sizes and usually produce useful luminous flames with large radiating surfaces. Their design is very complex but there are no moving parts and in operation they are simple and easily maintained. All the air used comes from the atmosphere outside the burner and provision must be made for its adequate supply preferably through adjacent vents. (See Fig. 14(c)).

Where the gas volume rate is high, combustion takes place over a considerable space and clearly some time is required to permit complete combustion to take place. Particularly in the case of a lean gas like producer gas, but also when, say a town gas is being burned, "burners" may be very simple and unsophisticated, being little more than open ends of pipes delivering the gas into the furnace alongside larger ducts carrying in preheated air. These entries would be called gas and air "ports" and they may be of such dimensions and so disposed as to induce a useful amount of turbulence into the gas and air streams. The flames produced are called "diffusion flames" but the name is scarcely appropriate because the rate at which the reactions proceed depends on the amount of turbulence. Parallel streams of gas and air would interact rather slowly across a narrow combustion front which moves away from the gas or the air—whichever is in excess.

In most practical cases the flow is turbulent and ordinary molecular diffusion is swamped by the turbulence. This turbulence is increased by:

(a) A difference in the velocities of the gas and air streams.
(b) A difference in the directions of the gas and air streams.
(c) A difference in densities of the gas and air streams, if the heavy one is on top.

The greater these differences the faster the reactions and the hotter the flame. An excess of air accelerates combustion as does a further increase in turbulence by baffles. The use of oxygen enriched air or pure oxygen would be expected to increase reaction rates and would of course increase the flame temperature for other reasons too. The combustion front would extend from the burner inclined away from the side of the excess air or gas.

These general considerations have been applied in the design of the furnace burners or "ports" in the open hearth steel furnace as it developed for the burning of gas. The angle and dimensions of the gas and air ports were critical but the result was a furnace lacking in versatility. Attempts to introduce variable operation were never mechanically acceptable and a few well tried designs such as the Venturi furnace became almost standard. With the modern trend toward oil firing it seems that no further development of this type of furnace is likely at present. Nevertheless the same principles apply to many other kinds of furnace in which gas will continue to be used, and the means of designing for short hot flames or long cool flames with uniform heating capacity over a wide area are well known.

Reactions in the Flame

The chemical reactions of combustion have all been presented in terms of the familiar chemical species of the known reactants and products. Investigations into the kinetics of these reactions particularly in the gaseous phase show that other molecules or free radicles are present in flames. These include the atoms O and H and other groups like OH, HO_2, HCO, CH, and C_2, and when oil is burning a wide variety of organic groups have been detected also, mainly by emission and absorption spectroscopy. The presence of these transient molecules indicates that the reactions represented by the simple equations probably occur in series of stages involving some or all of these groups. This is compatible with the modern activated complex theory of reaction mechanisms and leads to interesting speculations as to the actual sequences which occur and on the chain reactions which would best explain the extreme speed of some of these reactions under suitable conditions.

Controlled Atmospheres

It is sometimes necessary to control the atmosphere in a furnace so that it will not react with the stock in an undesirable way. The commonest example is the maintenance of slightly reducing conditions over the metal surface in steel heat-treatment furnaces. In these cases the chemical considerations must be given priority over fuel efficiency and burners must be designed to operate with a slight deficiency of oxygen to produce the gases with some carbon monoxide. Otherwise H_2–N_2 mixtures may be used in sealed off muffles, or salt baths may be used to prevent atmospheric attack. Another way is to use so-called "endothermic" gas prepared by partial oxidation of town's gas, or other fuel gas, under close control in a catalytic reactor. The product may be freeze-dried and used at any selected reduction potential, i.e. CO/CO_2 ratio.

12

The Conversion of Electrical Energy to Heat

THERE are four types of electric furnace:

1. Resistance furnaces.
2. Arc furnaces.
3. Induction furnaces.
4. Capacitance furnaces.

Of these the first three are important to metallurgists.

1. *In Resistance Heating* a current is passed through a conductor (or resistor). The energy dissipated is given by:

$$\text{Energy in watt-seconds} = V \times I \times t \qquad (12.1)$$
$$\text{(Joules)} = I^2 \times R \times t \qquad (12.2)$$

where

V is the voltage drop across the resistor;

I is the current flowing in amperes;

R is the resistance in ohms;

and

t is the time in seconds.

If the supply is a.c., V and I are r.m.s. values and the formula assumes a power factor of 1. This may be re-written:

$$\text{Energy in kcal} = 0 \cdot 000239 \, I^2 R t \qquad (12.3)$$

assuming that all the energy goes to heat which is very nearly true.

139

A large number of materials are available for use as resistors and the properties of some of them are indicated in Table 6. These are divided into those with high and low specific resistance. High values are obviously an advantage, but in practice the use of low voltage and high amperage is usually found necessary in large furnaces even with nichrome and kanthal. This is because

TABLE 6

Materials Available as Resistors

Type	Specific resistivity	Melting point °C	Oxidation resistance etc.
Nichrome	High	about 1500	good to 1100°C
Kanthal	High	about 1500	good to 1250°C
Carbon	Low	none	nil—use reducing atmosphere or vacuum
Silicon Carbide	Low	none	good to 1550°C
Molybdenum Silicide	Low	none	good to 1700°C
Molybdenum	Low	2600	nil; use H_2 or vacuum
Tungsten	Low	3400	nil; use H_2 or vacuum
Tantalum	Low	2850	nil; use H_2 or vacuum
Platinum and its alloys	Low	1775	good
Salt Mixtures	Medium	from 250	good but salts volatilize and decompose

thick wire must be used in order that the surface area of wire shall be big enough to dissipate the heat at the required rate, and the wire be reasonably strong at the working temperature.

Most metal wires oxidize at medium high temperatures and some of them embrittle with continued use. The industrial use of molybdenum and tungsten is restricted to occasions when a hydrogen atmosphere is not intolerable, and carbon can be used only in an atmosphere of carbon monoxide. Platinum alloys are good to beyond 1600°C, or pure sintered rhodium up to about 1850°C, but these are very expensive.

Obviously, resistance heating is limited to moderately high temperatures only up to about 1250°C except for special applications. Up to furnace temperatures of 1500°C silicon carbide is the most suitable resistor material for industrial use. It is rigid even at high temperatures, but brittle. Oxidation is retarded by a silica film on the surface, and if this is not maintained a violent burn-out may occur. The resistance of silicon carbide elements increases slowly with age.

Molybdenum disilicide (Mosilit, Superkanthal) can be used up to a furnace temperature of about 1650°C. It too is protected from the atmosphere by a film of silica. It does not "age" like silicon carbide but maintains constant resistance over long periods though it becomes very brittle. These elements are a recent development and are very costly.

The platinum metals and their alloys—especially those of platinum and rhodium—are useful beyond 1600°C and fail usually by volatilization. They become brittle and coarse grained in reducing atmospheres particularly if silicon monoxide is present —or silica and a reducing agent. The very high cost of these metals discourages their use in industrial furnaces but they are widely used in laboratories.

The nichrome–kanthal groups are limited by failure of oxidation resistance at temperatures which vary from 1100 to 1250°C depending on type. Those containing aluminium have superior oxidation resistance but are more readily embrittled. Elements require to be well supported on well designed refractory mounts as their strength at elevated temperatures is low. These are the commonest elements for industrial resistance furnaces.

Carbon is the most refractory resistor material and is used as rods, tubes, slotted tubes and granules in various forms of furnace, capable of attaining over 3000°C under reducing or vacuum conditions. The main application is to laboratory and test equipment but occasional larger scale applications do occur.

Salt baths may be heated by passing an electric current between electrodes immersed in the salt. This type of furnace usually operates on very high current, at low voltage. It is limited in temperature range by the properties of the salts used which may be nitrates which can be explosive if raised to too high a temperature (550°C), or chlorides which can be used at higher temperatures but soon become volatile (e.g. 800°C). The extension of this system to higher temperatures might be effected by the use of silicates, borates, etc., but refractory difficulties would soon be met.

In certain electrolytic processes such as the extraction of aluminium by electrolysis of bauxite dissolved in molten cryolite, a high temperature is maintained by the current passing through the cell (Fig. 57).

There is nothing quite equivalent to theoretical flame temperature in electrical resistance heating, unless it is the maximum temperature at which the resistor can be operated. Watts dissipated in any section of the resistor must be equivalent to the heat lost by that section to its surroundings. If heat is lost entirely by radiation, and assuming black body conditions for simplicity:

$$I^2R = A\sigma(T_E^4 - T_F^4) \quad \text{(see p. 164)} \tag{12.4}$$

where T_E and T_F are the temperatures of the element and the furnace respectively, and A is the effective radiating area. During heating up from cold, if I^2R is kept constant T_E must rise as T_F rises. Ultimately an equilibrium will be attained at which:

$$I^2R = A\sigma(T_E^4 - T_F^4) = \text{heat lost by the furnace to} \tag{12.5}$$
$$\text{the surroundings.}$$

If the furnace were perfectly insulated T_F and T_E would increase together (T_E always ahead) until the element failed. The watts input to an element is limited by the capacity of its surface to emit heat at its maximum operating temperature. In furnace design, elements must be chosen to have enough surface area to transmit energy at the required rate. For this reason tape is sometimes preferred to round wire.

Elements are not usually straight, but coiled, so that part of the heat radiated is reflected within the coil and for any watts input the element runs at a higher temperature than if it were straight (other things remaining unaltered). In other words, any particular element temperature can be maintained with a lower power consumption or, if the temperature *and* the power are maintained constant a greater length of resistors can be incorporated into the furnace design and a higher input becomes possible when required. This might be expressed that the electrical equivalent of flame temperature can be raised by using coiled elements.

Heating elements usually fail by loss of section due to oxidation, volatilization or creep. If one part becomes rather thinner than the rest, the resistance rises locally but the current remains substantially constant. The watts dissipated per unit length rises at the thin part; the temperature rises; and at once the rate of oxidation, volatilization or creep increases. A "hot spot" has developed and failure is inevitable. The other main cause of failure is fracture, following embrittlement, under vibrational or thermal stress.

Resistance heating in all forms is common in laboratories where it is convenient, readily controlled and clean, but its cost is rather high for industrial applications unless some distinct advantage is to be gained over other forms of heating. These advantages would be mainly in control, in uniformity of heating, and in freedom from combustion gas atmosphere. The main metallurgical applications are in heat-treatment furnaces where controlled atmospheres are often desirable and occasionally for melting low melting point metals. It is also used in forced air circulation furnaces designed for uniform heating at low temperatures up to 700°C (Fig. 56).

2. *Electric Arc* heating is a special form of resistance heating in which the resistor is the "plasma" or ionized atmosphere created in the space between two electrodes. The mass of this resistor is very small but it is indestructable and operates at an extremely high temperature, about 5000°C. The major applications should be where the "flame temperature" required cannot be attained

by other means. The greatest use is in the manufacture of special steels but it is also used in many smelting processes and for the preparation of sound ingots especially of refractory metals like titanium by vacuum arc melting—a process rather like electric welding in an evacuated chamber. Large ingots of steel are also made by remelting selected material in this way. Electric arc welding is obviously a further application of arc heating.

The current used may be a.c. or d.c., and if a.c., single or poly-phase supply. In steelmaking three-phase a.c. is the rule, using consumable carbon electrodes (Fig. 44). Consumption of power is very high—50,000 A at 500 V—and the load fluctuates violently especially during melting down. A variable reactor—a sort of choke—must be incorporated in the circuit to suppress violent surges, and robust circuit breakers to isolate the unit from the high tension supply when on overload. The gap between the electrodes (or between the electrodes and the load if that is part of the circuit) is kept controlled toward a constant resistance during operation. This is particularly important during melting down.

A power factor of about 0·9 can be maintained without reactance if the maximum operating power is controlled to an optimum value which is somewhat under the maximum possible.

3. *Induction Heating* is again a special form of resistance heating, the resistor this time being the burden itself in which an electric current is induced. The principle is that of the transformer, the supply being conducted round a primary circuit so disposed that the secondary current flows in short circuited turns within the burden.

There are two important types of furnace—the cored type and especially the Ajax–Wyatt furnace (Fig. 30) and the coreless type typified by the Leeds–Northrup furnace (Fig. 31).

In the cored type the transformer pattern is quite obvious (Fig. 15(*a*)). The primary current is passed through a coil housed in the hearth of the furnace. Laminated iron cores concentrate the electro-magnetic field so that an efficient coupling is effected with the narrow loop of molten metal which passes from the crucible underneath the primary coils and, concentric with them, back to the crucible. This loop of molten metal carries the secondary current and acts as a resistor. The metal in the loop is driven by motor action and sets up a circulation of the contents of the crucible so transferring the heat evolved through the melt. This type operates at normal frequency (50 Hz). It is simple in design and operation but in so far as the loop must be maintained molten, the crucible can be used only for one type of alloy and

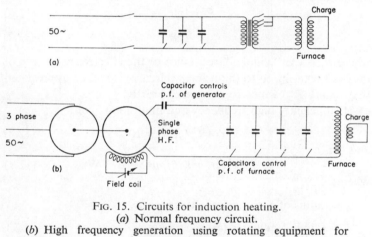

FIG. 15. Circuits for induction heating.
(a) Normal frequency circuit.
(b) High frequency generation using rotating equipment for 2000–10,000 Hz.

must be kept in continuous production. The dimensions of the loop are designed to suit the properties of the alloy being melted. Erosion of the refractories enclosing the loop limit the scope of this type of furnace. It is used very successfully with brass and other copper base alloys but not with ferrous alloys which require a much higher temperature. Aluminium alloys deposit oxide which chokes the loop and modified designs have been developed for them.

The coreless induction furnace is of simpler construction but operates at higher than grid frequencies and requires more elaborate electrical equipment (Fig. 15(b)). For melting, the crucible is situated within a water cooled helix which carries the primary current and heat is generated within the charge by induced eddy currents.

The underlying theory, when fully developed, is complex and design is in part empirical. The salient points are summarized here.

If I_1 and I_2 are the currents flowing in the primary and secondary circuits and R_2 the resistance in the secondary, the heat developed in the load is given by:

$$P = I_2^2 R_2 \qquad (12.6)$$

$$= I_1^2 R_c \qquad (12.7)$$

where R_c is the "coupled" resistance or the effective resistance of the load transferred to the primary circuit. It will be appreciated that I_2 and R_2 cannot be measured directly.

At sufficiently high frequency and in a "long" coil:

$$R_c = 4\pi^2 r_0 n^2 l \sqrt{(f\rho\mu)} \times 10^{-9} \ \Omega \qquad (12.8)$$

and hence

$$P = 4\pi^2 I_1^2 r_0 n^2 l \sqrt{(f\rho\mu)} \times 10^{-9} \ \text{W} \qquad (12.9)$$

where:

$r_0 =$ radius of a cylindrical charge (cm)

$f =$ supply frequency (Hz)

$\rho =$ resistivity of charge ($\Omega-$cm $\times 10^{-9}$)

$\mu =$ "effective" permeability of charge (unity for non-magnetic material and perhaps 50 or 100 or more for magnetic material, depending on flux density, etc.)

$n =$ turns of coil per cm

$l =$ length of charge (cm)

In the practical case this value of P will be diminished by end effects in relatively short and wide coils and by the dimensions and shape of the charge.

Induced alternating current flows in the surface of a conductor to an effective depth

$$p = \frac{1}{2\pi}\sqrt{\frac{\rho}{\mu f}}\,\text{cm} \qquad (12.10)$$

about 90 per cent of the heat being generated in a surface layer of this thickness. The layer is obviously thinner for high frequency current than for low, and thinner in ferromagnetic material than in non-magnetic. Typical values of p would be, at 5000 Hz 0·3 mm in cold ferromagnetic steel, or 7 mm in the same material above the Curie point, while at 500 Hz the values would be about 1 mm and 20 mm respectively. It is obvious that it would be an advantage if the diameter of the charge were $2p$ cm when possible, and in practice large numbers of small pieces are charged to melting furnaces and each piece heats up separately and quickly, but soon welding occurs and ferromagnetism (if any) is lost at the Curie point. Later the charge melts and with r_0, ρ, μ and the dimensions and shape of the charge all altering, R_2 is reduced, and P is also reduced unless there is provision either to increase the applied voltage and hence I_1^2, or to increase f.

An increase in f would concentrate the heating effect in the outer layers of the melt and reduce the stirring action of the eddy currents. At this stage a relatively low frequency would be most appropriate with I_1^2 increasing, the control being effected at the field coil of the generator. In fact control by varying frequency could not easily be applied.

Inductance in the coil causes considerable lag and the power factor will drop to about 0·5 with a magnetic charge, or to 0·1 or less above the Curie point. A variable bank of capacitors (Fig. 15) is used to maintain the power factor near unity—0·9 lag being an accepted lower limit. This accounts for a large part of the cost of the equipment.

Coreless induction furnaces are built up to about 20 tons capacity and 2000 kVA rating operating at frequencies up to about 5000 Hz (lower in bigger sizes). The voltages at the coil would be about 750 in the larger sizes. Energy losses would include 10–15 per cent copper losses (to the cooling water), 2–3 per cent losses in the capacitors and 20–25 per cent thermal losses. A fairly high over-all efficiency depends upon rapid and continuous operation. These furnaces remain fairly simple in design and there would appear to be room for development perhaps toward two-stage operation—(1) preheating and (2) melting and super-heating of the molten charge—the electrical arrangements in each stage being chosen to suit the condition of the charge.

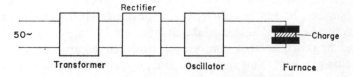

FIG. 16. Circuit for capacitance heating.

The metallurgical advantage of induction melting lies in its cleanliness coupled with speed (in a small unit) and the possibility of employing an inert atmosphere or even vacuum. Slag refining is not easy unless stirring-in the slag using low frequency eddy currents can be tolerated, because otherwise the slag layer on top is rather cold so that slag metal reactions go very slowly. The main applications are to special steels and other high melting point quality alloys like nimonics, where the process is little more than melting and pouring. Similar equipment has also been used for heating billets for forging. It lends itself to production line conditions and gives good reproducibility of temperature distribution and minimum scaling of the surface, because of the rapidity of the heating.

Induction heating for heat treatment and particularly surface hardening permits very rapid heating with localized effect if desired. The job may be within the coil or a flat coil may be used external to the part to be treated. The frequency may be very high (10^5–10^6 Hz) for surface heating. Other applications include brazing, welding and sintering and there are many laboratory uses including the vacuum melting of reactive metals like titanium suspended in space under the action of an electromagnetic field (levitation melting). Casting into a waiting mould is effected by reducing the field strength and letting gravity take over.

4. *Dielectric or Capacitance Heating* can be applied only to non-conductors. The load is placed between two electrodes forming a condenser in a circuit such as Fig. 16. Here it is subjected to rapidly alternating strain and heats by "molecular" friction much as does a metal rod subjected to alternate bending. Since the power developed $P = f.I.V^2$ a high power requirement can be attained by increasing V or f. High voltage would lead to arcing but frequencies up to 200 MHz can be employed for very rapid heating of materials in, for example, plastics and ceramics industries and for cooking. Heat is developed simultaneously at all depths and not only at the surface. There are no important metallurgical applications so far.

13

Heat Transfer

THE liberation of heat energy is only the first stage in the heating operation. This energy has to be transferred to the stock to be heated as completely as possible and its dissipation to the surroundings reduced to a minimum.

Heat transfer occurs by three mechanisms which usually operate simultaneously but often with one predominating—these are conduction, convection and radiation.

Conduction is transfer of heat by mutual interactions of vibrating atoms which themselves maintain their mean positions unchanged. Obviously this is most important in solids where the mobility of the atoms is least. While conductivities are quoted for both liquids and gases they can be measured only under rather restricted conditions.

Convection is transfer of heat due to the motion of atoms, molecules or aggregates of molecules carrying heat from one place to another. Convective transfer occurs in liquids and gases, and may be natural or forced. Natural convection is usually caused by the buoyancy of the hottest fluid causing it to rise against colder fluid so setting up familiar "convection currents". If the fluid flow is imposed by external forces the resultant heat flow is by forced convection.

When a solid surface is surrounded by liquid or gas at a lower temperature the fluid heats up at the solid surface and convection currents are set up which cool the surface. If a fan or impeller is placed in front of the surface it will cool more quickly—by forced

convection. In either case the actual heat transfer from the solid
to the fluid is probably by conduction at the interface across a
thin (~ 0.05 mm) layer of fluid associated with the surface and in
which any flow is parallel with the surface, and then by convection
into the moving fluid by its interaction with this thin film.

Radiation is heat transfer through space as photons of electromagnetic radiation of wave lengths greater than 10,000Å. All
solid surfaces above 0°K radiate and absorb heat to and from their
surroundings so that it is only net heat transfer that is important.
Radiant heat may be reflected or transmitted or absorbed like
light, and many gases and liquids are transparent to it.

Quantitative aspects of heat transfer will now be dealt with
in some detail.

Conduction

Conduction of heat in solids is described by the classical
equation:

$$dH = K.dA.dt.\frac{dT}{dx} \tag{13.1}$$

where dH is the quantity of heat flowing across an area dA in
time dt down a temperature gradient dT/dx. This well known
equation is analogous to those for electrical conductivity and
mass diffusivity. The constant K is the thermal conductivity of
the material. K depends on temperature and also on the physical
condition of the material particularly its porosity if it is or a
porous or granular nature. It is not easily determined because
of the experimental difficulty of arranging parallel heat flow, and
really accurate values of conductivities of ordinary materials over
wide temperature ranges are not readily available.

It is usually difficult to apply the equation to the determination
of heat flow rate and temperature distribution, in practical cases,
for reasons of geometry and because K varies with T. Also unless
heat flow is substantially parallel a more general relationship

must be used:

$$K\left(\frac{d^2T}{dx^2} + \frac{d^2T}{dy^2} + \frac{d^2T}{dz^2}\right) = C\frac{dT}{dt} \qquad (13.2)$$

where T is the temperature at point (x, y, z) at time t and C is the specific heat per unit volume. The ratio K/C is called thermal diffusivity. The use of this equation involves integration and selection of suitable boundary conditions. Usually, either the boundary conditions chosen involve unjustifiable simplifications or the mathematics becomes impossible. Increasing availability of computers, however, steadily increases the range of problems that can satisfactorily be tackled.

Under steady state conditions, however, and where the geometry is simple it is quite easy to calculate heat through-put for a furnace wall, for example, by a calculation exactly analogous to an application of Ohm's Law. The assumption must be reasonable that the wall is an infinitely extending slab of finite thickness d_1. If the thermal conductivity is K_1 the thermal resistance is then d_1/K_1 per unit area. The potential is the temperature difference across the slab and the heat flow per unit area corresponds to electrical current. Equation (13.1) can then be applied in the same way as Ohm's Law might be used, to determine any one unknown value. If a wall were made up of several materials, different kinds of brick for example, the composite thermal resistance would be the sum of the several resistances:

$$R = \frac{d_1}{K_1} + \frac{d_2}{K_2} + \frac{d_3}{K_3} \qquad \cdot (13.3)$$

and other terms might be incorporated to represent the thermal resistance of, say, cement or an air space between the various sections. Again eqn. (13.1) will solve for the heat flow rate if required, K/dx being replaced by $1/R$. In a case like this, once dH/dA has been determined the several temperatures at the interfaces between the different kinds of brick can be calculated by further application of eqn. (13.1) to each section of the wall in turn. In the practical case of finite walls some adjustment would have to be made for ends and corners and other geometric features like doors, possibly using a "relaxation" method invoking continuity of the isotherms and the rationalization of their spacing and that of the heat flux lines orthogonal to them.

Another case where a reasonable estimate of steady state heat flow can be made is to the hollow cylinders representing a steam pipe or hot gas main. If the length is l, the internal and external radii R_1 and R_2 and the temperatures of the inner and outer surfaces T_1 and T_2 the heat flow can be shown to be

$$dH = \frac{2\pi l K(T_1 - T_2)}{\ln R_2/R_1} \tag{13.4}$$

If the wall of the cylinder is thin ($R_2 < 2R_1$) this can be reduced to

$$dH = \frac{\pi l K(R_1 + R_2)(T_1 - T_2)}{(R_2 - R_1)} \tag{13.5}$$

which corresponds closely to eqn. (13.1), the wall being treated as a slab.

If the cylinder wall is composite—for example, brick lining—steel shell—lagging—the same device of adding the resistances is employed and eqn. (5.4) is extended to

$$dH = \frac{2\pi l(T_1 - T_2)}{\dfrac{\ln(R_2/R_1)}{K_1} + \dfrac{\ln(R_3/R_2)}{K_2} + \dfrac{\ln(R_4/R_3)}{K_3}} \tag{13.6}$$

where R_1, R_2, R_3, etc., now represent the internal radii of the several layers and K_1, K_2, K_3, their conductivities. Similar treatment might obviously be extended to deal with oval ducts but if applied even to relatively simple rectangular sections it would not yield a very satisfactory solution. The previous treatment as for slabs, with corrections for corner effects would probably be better in that case.

These formulae could be applied as they stand only to lengths of cylinder in which the temperatures of the inner and outer surfaces were substantially constant. If the temperatures do vary along the duct, either mean temperatures have to be assessed, or the duct considered in a number of sections, each of which satisfies the above condition and can be worked on separately.

Where geometric complexity is too great for mathematical treatment the electrical analogue is the most convenient device for assessing heat transfer. A network of resistances is constructed to simulate the assembly of sections of the object being studied. Electrical potentials are applied to represent temperatures where these are known and potential readings taken elsewhere as a guide to temperature distribution, while ammeters inserted in the network indicate heat flow rates. Variations of K with temperature can be incorporated, and improved estimates made as the problem is resolved. A similar analogue can be set up using the flow of a liquid in tubes but this is less convenient.

Non-steady state conditions, even when the geometry is simple also require difficult mathematical operations. Such problems concern the rate at which heat progresses into an object placed in a hot furnace or through the walls of a furnace during the heating up stage or during cooling. It is possible to calculate the rate of advance of the isotherms in the semi-infinitely thick mass, with a constant surface temperature and to show that this is inversely proportional to the distance from the surface and that the temperature distribution depends on the dimensionless CL^2/Kt where L is the depth from the surface. This may give a guide to heating rates in practical cases and will help too, to convert knowledge of one material to an estimate of the behaviour of another with different C and K. Speirs[4] describes two fairly simple methods of handling this type of problem. One is applicable to simple shapes and uses simple dimensionless groups (see next section). Prepared empirical graphs are necessary to effect a solution and the reader is referred to Speirs or to the original paper* for details. The same author also describes the Schmidt method which can be applied to heat flow through a slab "suddenly" brought to a

* RUSSELL, T. F., Iron and Steel Institute Special Report 14 (1936), p. 169.

high temperature on one face. A standard time interval is calculated peculiar to each problem and the temperatures at a series of equally spaced distances from the hot face are calculated by a simple graphical operation, for times corresponding to one, two, three, etc., of these time intervals. At present, however, the best way of finding out how long a particular article will take to heat to the centre under particular conditions is probably to carry out an experiment if this is at all possible.

Table 7 indicates the values of K for some common materials. Among the gases hydrogen and helium are outstandingly good

TABLE 7

Thermal Conductivities of Various Materials—in c.g.s. units*

Gases	Hydrogen	At room temperature	0·00040
		At 1000°C	0·00107
	Helium	At room temperature	0·00034
	Air	At room temperature	0·000058
	Carbon monoxide	At room temperature	0·000055
	Carbon dioxide	At room temperature	0·000035
	Methane	At room temperature	0·000072
	Water Vapour	At room temperature	0·000036
		At 1000°C	0·000126
Liquids	Water	At room temperature	0·0013
	Oil	At room temperature	0·0004
	Mercury	At room temperature	0·02
	Aluminium (Liquid)	At 700°C	0·02
Metals	Aluminium	At room temperature	0·57
	Copper	At room temperature	0·91
	Silver	At room temperature	1·0
	Iron	At room temperature	0·18
Non-metals	Carbon	At room temperature	0·1
	Firebrick	At room temperature	0·002
	Insulating Firebrick	At room temperature	0·0007
	Kieselguhr	At room temperature	0·0001
	Asbestos	At room temperature	0·00008

* SI: To convert to W/mK multiply by 418·68.

conductors. K in gases depends on the mean free path of the molecule and therefore increases with the temperature. The distinction between metals and non-metals is obvious and it is also clear that liquids have generally lower conductivities than solids. Data for liquids at high temperatures is virtually unobtainable.

Natural Convection

The mathematical treatment of convective heat transfer is even less satisfactory than that of conduction. A common problem is the estimation of heat flow rate at gas–solid interfaces. This depends on the thickness and thermal conductivity of a layer of gas presumed to be attached to the solid surface and to move relative to it in a laminar manner (Fig. 17). Under conditions of streamlined flow a series of parallel layers of gas are envisaged as sliding past a static layer attached to the surface, and each other with velocities increasing as this distance from the surface increases. Any heat passing normal to the surface would then be transferred only by conduction. Obviously heat flow is partly dependent on fluid flow and the various properties of the fluid relevant to flow must be considered.

In the case of natural convection, where the flow is caused by density differences between hot and cold fluid, the coefficient of thermal expansion, the specific heat and the viscosity of the fluid are all involved. With so many variables in the system it is obviously appropriate to use the mathematical theory of similarity to handle these problems just as it is used in fluid flow problems (Chapter 15).

In the theory of similarity or of dimensionless groups it is presumed that the behaviour of any physical system can be described through the relationships between the values of carefully chosen "dimensionless criteria", that is, groups of values of ordinary properties raised to appropriate powers and multiplied together so that the product has zero dimensions. For complete treatment of a system one suitable group should be found for every physical factor in the problem, and it is of course necessary to find relationships between the groups, taken two or three or more at a time so that soluble equations emerge for the unknowns when all the known values are inserted. These relationships should remain

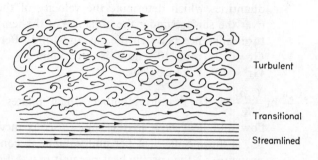

Fig. 17. Transition from streamlined to turbulent flow at a solid surface.

valid over wide ranges of conditions—for example, should cover convective transfer both by gases and liquids. In the case of natural convection the factors determining the heat flow H per unit area in unit time from a solid surface are:

a—the thermal expansion coefficient for the fluid,

g—the acceleration due to gravity,

θ—the gas–solid temperature difference,

D—a characteristic dimension of the system such as the diameter of a pipe,

C_p—the specific heat of the fluid per unit mass,

K—the thermal conductivity of the fluid,

$v = \eta/\rho$—the kinematic viscosity of the fluid, where η is its absolute viscosity and ρ is its density.

All these quantities must be expressed in self-consistent units.

They give rise to three relevant dimensionless groups—

$\text{Nu} = \dfrac{HD}{K\theta}$ is the Nusselt Number and is the criterion containing H

$\text{Gr} = \dfrac{ag\theta D^3}{\nu^2}$ is the Grashof Number which involves all the quantities which determine the velocity of the gas over the surface under the influence of the temperature difference between the gas and solid. Between $\text{Gr} = 10^3$ and 10^9 flow is streamlined. Above $\text{Gr} = 10^9$ it is turbulent.

$\text{Pr} = \dfrac{_vC_p\nu}{K}$ or $\dfrac{C_p\rho\nu}{K}$ is the Prandtl Number which relates fluid flow and heat flow in the same viscous medium. $_vC_p$ is the specific heat per unit volume at constant pressure, C_p the specific heat per unit mass. For all diatomic gases and for carbon dioxide $\text{Pr} = 0\cdot74$; for water vapour it is rather higher but for ordinary furnace gases at atmospheric pressure can usually be considered constant at the above figure. Steam or liquids would require another figure of course.

These three dimensionless groups are related by the following empirical formulae*—

for streamlined flow

$$\text{Nu} = C_S(\text{Gr}.\text{Pr})^{0\cdot25} \qquad (13.7)$$

and for turbulent flow

$$\text{Nu} = C_T(\text{Gr}.\text{Pr})^{0\cdot33} \qquad (13.8)$$

The values of C and the exponents were obtained from plots of log Nu vs. log(Gr.Pr) obtained from experiments in which convective heat transfer was measured under a very wide range of conditions and including both liquids and gases. The geometry of the system had to be maintained throughout a series of experiments, however, so that plane vertical surfaces would be

* LANDER, T. H., *Proc. Inst. Mech. Eng.* **148–9**, p. 81 (1942).

studied in one series, horizontal in another, horizontal cylinders,
fine wires etc., in other series. The form of the curves obtained in
these experiments are all rather similar (Fig. 18) with straight

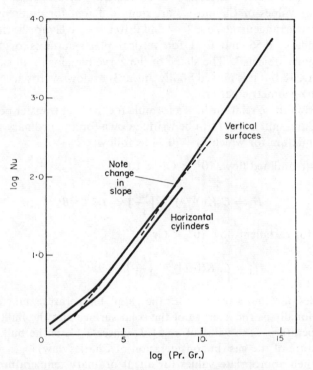

FIG. 18. Observed relationships between Nu and (Pr.Gr) from
which the coefficients and exponents of eqns. (13.7) and (13.8) can
be derived.

portions between log(Gr.Pr) values of 3 to about 8 having slopes
of 0·25 and another straight section developing beyond log
(Gr.Pr) = 9 which has a slope of 0·33. The initiation of turbulence has been observed by optical means at Gr = 2 × 10⁹. It
has been suggested that under fully turbulent conditions the

exponent of Pr should be 2 so that the viscosity term would disappear but this has not been observed under any conditions examined, presumably confirming the persistence of a thin surface film in which flow is laminar at all velocities investigated.

The values of C_S and C_T are quite different for different geometric arrangements, e.g. 0·47 and 0·10 respectively for horizontal cylinders; 0·56 and 0·12 for vertical plane surfaces or "large irregular objects". The shape of the curve remains in all cases as in Fig. 18 but it is moved bodily upwards or downwards according to the geometric pattern.

From these relationships a formula for the heat transfer per unit area and unit time H can be written down for any ordinary gas or gas mixture for which Pr = 0·74 as follows:

for streamlined flow, $10^3 < Gr < 10^9$

$$H_S = C_S K (0·74)^{0·25} \left(\frac{ag}{v^2}\right)^{0·25} D^{-0·25} \theta^{1·25} \qquad (13.9)$$

and for turbulent flow $10^9 < Gr$

$$H_T = C_T K (0·74)^{0·33} \left(\frac{ag}{v^2}\right)^{0·33} \theta^{1·33} \qquad (13.10)$$

Values of K and v are those for the "film" temperature, taken conventionally as the average of the solid surface and the bulk fluid temperature. The value of a is $1/T_g$ where T_g°K is the bulk temperature of the gas (in consideration of Charles' law).

When appropriate values for air at ordinary temperatures are inserted along with the coefficients for vertical surfaces or large irregular bodies, these formulae are reduced to

$$H_S = 1·03 \times 10^{-4} D^{-0·25} \theta^{1·25} \text{ cal/cm}^2 \text{ sec} \qquad (13.11)$$

and

$$H_T = 2·7 \times 10^{-5} \theta^{1·33} \text{ cal/cm}^2 \text{ sec} \qquad (13.12)$$

which could be applied to the natural convective cooling of, for example, ingots or ingot moulds, or furnace exteriors (ignoring other modes of heat transfer). In this case the product Gr.Pr has a value about $10^2 \theta D^3$. If θ is 100°C and D is 30 cm (1 ft) Gr.Pr is almost 10^8 and for any higher values particularly of D the turbulent conditions must hold. Most cases encountered in industry would therefore require the use of the eqn. (13.12).

Another formula is commonly employed to give the same information, namely

$$H_T = 4 \cdot 6 \times 10^{-5} \theta^{1 \cdot 25} \tag{13.13}$$

This is an approximation of (5.12) developed by Fishenden and Saunders[4] by incorporating the factor $\theta^{0 \cdot 08}$ in the "constant" term. It is valid up to about 800°C surface temperature but offers no advantage over (5.12). In either formula the coefficient for convection from an upward facing surface will be about 30 per cent higher and from a downward facing surface 35 per cent lower. Coefficients for large cylinders lying horizontal are about 20 per cent lower but those for narrow cylinders or wires are many times higher probably reflecting the different flow pattern produced as film thickness and wire diameter become comparable.

The solution of other problems involving other gases or liquids, other spatial arrangements and possibly heating of the surface rather than cooling of it can all be tackled in a similar manner. Relevant data for gases are readily available in references[4] and [19] but care must be taken with units. The accuracy of such calculations is not very high, and always depends on the results of someone else's experiments. Surface roughness, film thickness and actual temperature distribution are "averaged out" in the empirical formulae. Surface temperatures must be influenced by the convection, and not always in a uniform manner. Even the bulk temperature of the gas need not be uniform opposite all parts of a surface. In practice too, natural convection is often accompanied by some forced convection—stray draughts for example—so that observed heat transfer is often greater than that calculated by these formulae, assuming natural convection only.

Forced Convection

Forced convection may be superimposed on natural convection and may easily be so effective as to swamp it completely. Like natural convective transfer, it is much greater under turbulent flow than under laminar flow conditions and the velocity of the fluid over the solid surface is the most important single factor involved. If flow is laminar heat transfer normal to the surface will be by conduction through the fluid as far as the thermal gradient persists but under turbulent conditions conduction of heat normal to the solid surface is only across a thin film—about 0·05 mm—and the fluid temperature beyond that is kept practically uniform by the turbulence, so that the effective temperature gradient across the film is as high as it can possibly be. Obviously where convective transfer is being designed high velocities for turbulent flow should be arranged—the higher the velocity the thinner the film and the greater the heat flow.

The relevant dimensionless groups are the Nusselt and Prandtl Numbers and the Reynolds Number $Re = VD/v$ where V is the velocity of the fluid and the other terms are as in the previous section. If $Re > 2000$ the flow is turbulent. In a 30 cm (1 ft) diameter tube the critical velocity for cold air would therefore be 9 cm/sec or 6·3 1/sec; for hot air at 1000°C this value would be 117 cm/sec. It will be appreciated that most industrial processes use gas velocities which are high enough to cause turbulence but these critical velocities are quite high and they need not always be surpassed. If convective transfer is to be important design should be for maximum turbulence coupled with large reacting surface area, making use of corrugations, vanes and baffles.

For fluid passing through a pipe the general relationship (assuming turbulence) is

$$Nu = 0·023(Re)^{0·8}(Pr)^{0·4} \qquad (13.14)$$

As before for relevant gases $Pr = 0·74$ so that

$$Nu = 0·02(Re)^{0·8} \qquad (13.15)$$

If the pipe diameter is D,

$$H = \frac{0 \cdot 02\text{K}}{D} (\text{Re})^{0 \cdot 8} \theta \qquad (13.16)$$

$$= \frac{0 \cdot 02\text{K}}{D} \cdot \left(\frac{VD}{v}\right)^{0 \cdot 8} \theta \qquad (13.17)$$

which can be evaluated if the properties of the gas at the bulk temperature and the gas velocity are known.

If the film conductance $h = H/\theta$ is evaluated, its inverse can be incorporated into eqn. (13.3) so that the thermal resistance of the film is added to the other thermal resistances in series. Equation (13.3) then becomes

$$R = \frac{1}{h_1} + \frac{d_1}{K_1} + \frac{d_2}{K_2} + \frac{d_3}{K_3} + \frac{1}{h_2} \qquad (13.18)$$

representing the thermal resistance of a composite wall with a thin gas film resisting heat flow on either side and extending eqn. (13.6) to—

$$H = \frac{2\pi l(T_1 - T_2)}{\dfrac{1}{h_1} + \dfrac{\ln R_2/R_1}{K_1} + \dfrac{\ln R_3/R_2}{K_2} + \dfrac{\ln R_4/R_3}{K_3} + \dfrac{1}{h_2}} \qquad (13.19)$$

which represents the heat flow through a composite pipe wall— say gas film–scale–metal–scale–gas film.

The formula for determining H (or h) on the outside of a tube is similar to (13.14) but with a change both in coefficient and exponent—

$$\text{Nu} = 0 \cdot 26(\text{Re})^{0 \cdot 6}(\text{Pr})^{0 \cdot 3} \qquad (13.20)$$

and for gases, as above,

$$H = \frac{0 \cdot 24\text{K}}{D} \left(\frac{VD}{v}\right)^{0 \cdot 6} \theta \qquad (13.21)$$

provided the gas flow is at right angles to the tube axis. If the gas flows at 45° to the tube axis the coefficient is reduced to 0·18 or to 0·12 if the flow is parallel to the tube.

If tubes are arranged in banks so that gas passes through one row, and then a second and a third, turbulence increases through the bank and the coefficient is raised to about 0·32 but is subject to further adjustment for the exact geometric arrangement and spacing of the tubes—higher for staggered array, and closer setting except where Reynolds number is already very high and further improvements difficult to attain. Details of these adjustments and of other formulae to fit other conditions are available in reference books[4, 19] along with graphical aids to calculations.

These formulae are, of course, all empirical and of limited accuracy but they are extremely useful for predicting the effect of changes in those operational factors which are included in the dimensionless groups.

Radiation

Heat transfer by radiation is in accord with Stefan's Law which states that the amount of heat H_1 radiated in unit time from unit area of the surface of a "black body" at $T_1°K$ is given by:

$$H_1 = \sigma T_1^4 \qquad (13.22)$$

σ ($= 1·37 \times 10^{-12}$ cal/cm^2.sec.deg$^4 = 5·73 \times 10^{-8}$ J/m^2 s K^4) being a universal constant depending only on the units in use. The intensity of this radiation leaving any point is a maximum in the direction normal to the surface at that point and diminishes to zero according to a simple cosine law, in a direction parallel to the surface (Lambert's Law).

Simultaneously, the surface receives radiant energy from its surroundings at T_2 equal to:

$$H_2 = \sigma T_2^4 \qquad (13.23)$$

so that the net heat flow, per unit area in unit time is:

$$\Delta H = H_1 - H_2 = \sigma(T_1^4 - T_2^4) \qquad (13.24)$$

Successful application of this formula necessitates adjustments for the bodies not being "black" and for geometric conditions.

In general, when radiation falls on the surface of a slab of thickness d a fraction R is reflected while $(1 - R)$ enters the surface and is absorbed as it passes through the material. The fraction absorbed by the thickness d is given by $(1 - R)(1 - e^{-\alpha d})$ where α is the coefficient of absorption, and $(1 - R)e^{-\alpha d}$ is transmitted to the boundary where it is further subdivided to $R(1 - R)e^{-\alpha d}$ reflected internally and $(1 - R)^2 e^{-\alpha d}$ escaping the slab. At the same time the body is radiating heat. That part emitted by an element dx thick and x below the surface is $\varepsilon . dx$ per unit area of which $\varepsilon . dx\, e^{-\alpha x}$ reaches the surface without absorption, and there a fraction R is reflected internally and $(1 - R)\varepsilon . dx\, e^{-\alpha x}$ is emitted. Integrating between $x = 0$ and $x = d$ the total normal emission is found to be $\varepsilon/\alpha(1 - R)(1 - e^{\alpha d})$.

In most solids α is so high that absorption is practically complete in a very short distance d, so that all radiation not reflected is absorbed and the expression for total emission per unit area becomes:

$$\frac{\varepsilon}{\alpha}(1 - R)$$

and if reflection is negligible it is further simplified to:

$$\frac{\varepsilon}{\alpha}$$

for a black body. According to Kirchoff's law, this ratio is the same for all substances (at any given temperature and wavelength) so that ε is greatest when α is greatest, i.e. when $\alpha = 1$ or when the body is non-reflecting and completely absorbing.

In practice, absolute emissivity ε is not used, but a quantity E called emissivity or emissive power, which is the ratio of the intensity of radiation emitted by a surface at any given temperature to that of a black body at the same temperature:

$$\text{i.e. } E = \frac{\dfrac{\varepsilon}{\alpha}(1 - R)(1 - e^{-\alpha d})}{\dfrac{\varepsilon}{\alpha}} = (1 - R)(1 - e^{-\alpha d}) \qquad (13.25)$$

Likewise the absorptive power is defined as:

$$A = (1 - R)(1 - e^{-\alpha d}) \tag{13.26}$$

and clearly:

$$E = A \tag{13.27}$$

In a perfect black body $R = 0$ and $\alpha = 1$ so that $E = A = 1$ while the values of E and A of non-black bodies are fractions.

From this it follows that for a non-black body at a temperature T_1 in surroundings at T_2 the net heat flow per unit surface area becomes:

$$\Delta H = E\sigma(T_1^4 - T_2^4) \tag{13.28}$$

a maximum value, obtained only if a negligible part of the emitted energy is received back by reflection from the surroundings. This is the case if the body concerned is small in a relatively large cavity. If the body almost fills the cavity so that its area is nearly that of the surroundings then the emissivity of the surroundings becomes important and a minimum value is obtained.

$$\Delta H_{min} = \frac{E_1 E_2}{E_1 + E_2 - E_1 . E_2} \sigma(T_1^4 - T_2^4) \tag{13.29}$$

E_2 being the emissive power of the surroundings. In most practical cases ΔH is a compromise between these and there are a number of empirical formulae and aids published to assist in the assessment of ΔH under specific geometric conditions[19].

Some typical values of E are listed in Table 8. Apart from carbon black and platinum black and finely ground pigments there are few materials with E approaching 1. In these cases, the high value may be due to fineness of division, as a hollow acts as a trap for radiation and the closest approach to true black body conditions is a small hole leading out of a relatively large cavity, preferably blackened inside. Such an aperture would be used as a standard for determining values of E or in pyrometry. Bricks and other ceramics have E about 0·5–0·8 while metals, particularly if polished or molten, have very low values indeed, R being high.

The discussion so far has ignored the wavelength of the radiation. Strictly, the arguments developed apply to only one wavelength at a time and the full spectrum should be embraced by integration but the result would be substantially the same, for solids.

TABLE 8

Emissivities of Various Materials

Firebrick	0·75
Silica	0·66
Magnesite	0·4
Lamp black	0·98
Polished aluminium	0·08
Platinum (at 1000°C)	0·16
Steel (black plate)	0·95
Steel (black plate) at 500°C	0·85
Steel (ground surface)	0·25

The spectral distribution of energy emitted by a black body at $T°K$ is defined by Wien's Radiation formula:

$$E_\lambda = C_1 \lambda^{-5} \exp(-C_2/\lambda T) \qquad (13.30)$$

where E_λ is the energy emitted per unit area in the waveband λ to $\lambda + \delta\lambda$ and C_1 and C_2 are usually constants. The spectra for solids at several temperatures are shown in Fig. 19, in which the areas under the curves are proportional to the total energies emitted per unit surface area. It is particularly noticeable that the proportion of the energy which is radiated as visible light is very small. That the colour changes from red toward orange and yellow as the temperature rises is also obvious. "Grey" bodies have similar spectral distributions but C_1 being less than for a black body the intensity at all wavelengths is reduced. In cases where C_1 varies with λ the radiation is "coloured" and the treatment of heat transfer given above would require adjustment. Notable among "coloured" radiators are certain gases. Most diatomic gases are transparent to radiant heat and emit no radiation, but triatomic gases and particularly carbon dioxide and water vapour in furnace gases have absorption bands as indicated in Fig. 19 and hence also emit energy in these wavebands.

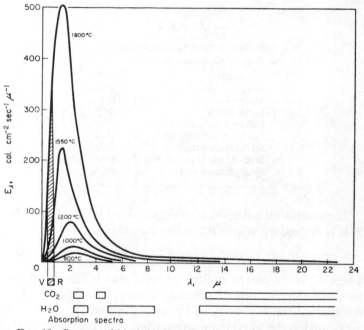

FIG. 19. Spectra of black body radiation at various temperatures. The visible range and absorption bands for CO_2 and H_2O are indicated below.

In so far as the absorption coefficient is low in these gases the thickness of the emitting layer of gas is important. A semi-infinitely thick layer will have $E = 1 - e^{-\alpha d} = 1$ and so radiate like a black body but only within the effective wavebands. Thinner layers will be like grey bodies within these bands. For Carbon dioxide a 1 m layer would have an emissive power about 0·2 at atmospheric pressure and a similar layer of water vapour about

0·4. Pressure of course also affects E and in exactly the same way as the effective layer thickness d. The relationship between pressure, layer thickness, temperature and emissive power are presented graphically in Fig. 20 for water vapour and carbon

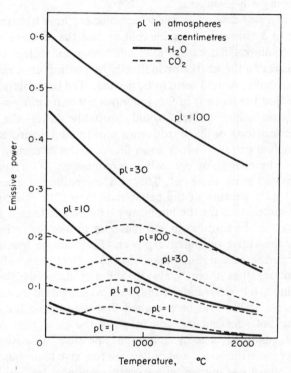

FIG. 20. The emissive power of CO_2 and H_2O each in a range of pressure (p) and/or effective layer thickness (l).

dioxide respectively, in an abridged version of working curves readily available in the literature[19] and originally determined by Hottel and Mangelsdorf.* In mixed gases the E values for the

* HOTTEL, H. C., and MANGELSDORF, H. G., *Trans. Amer. Inst. Chem. Eng.*, 31, p. 517 (1935).

components are not quite additive and the standard adjustments have to be made.[4] Further, the effective layer thickness depends on the shape of the flame[4] and is, for example, 1·8 times the actual thickness for a parallel faced sheet of gas, the "standard" shape being a hemisphere.

This can be very important where efficient heat transfer from a flame to a charge is desirable, but at best the emissive powers of non-luminous flames are low, and some convective transfer either direct to the charge or indirectly by heating up a roof and radiating from it would seem to be needed. The best solution is to arrange that the flame is in fact luminous either by using a slower burning technique (which would probably lower the flame temperature too) or by introducing unsaturated hydro-carbons (coal tar fuel or pitch) which form fine particles of carbon which impart luminosity except when combustion is particularly vigorous and burns them out. This also lowers the flame temperature but only because of the better radiation. The fine particles of char responsible for the luminosity may be less than 1 μ thick and may well be translucent but make up in numbers what they lack in absorptive or emissive power. They radiate much more heat than the best non-luminous flames. They may suffer the disadvantage that there is always a hotter flame which could be attained, but only under conditions which would destroy the luminosity. It has also been argued that they screen the charge from the hot roof but the advantage of the good heat transfer actually obtained usually outweighs possible disadvantages. Producer gas (from coal) and pulverized fuel give luminous flames as does oil—if not burned with oxygen—and coal tar fuel. Coke oven gas and natural gas however give transparent flames unless additions are made, and pitch is usually used for this purpose.

Practical Cases of Heat Transfer usually involve all three mechanisms and usually involve unsteady state conditions over a large part of the process. Heat flow rates by the different mechanisms would be calculated separately and the three values added. Conduction and convection may, as indicated, be treated together. It can happen that one mechanism predominates. The forced convective cooling of the surface of an excellent insulator would produce a thin chilled zone near the surface through which the conduction of heat would be slow—conduction soon being established as rate controlling. A steady state might eventually be set up and the physical laws would be obeyed, but K rather than V would determine the heat flow rate and the surface temperature would be far from that of the body interior. At very high temperatures radiation is very much more effective across space than convection to the extent that the latter may sometimes be ignored. Such a condition when recognized simplifies arithmetic.

At intermediate temperatures, where convective transfer is most likely to predominate, forced circulation should be applied wherever possible and every step taken to increase the gas velocity across the surfaces to be heated. Forced circulation furnaces (Fig. 56) are designed to this end.

Heat flows through a granular material, porous material, or beds of broken solids by a combination of conductivity through and between solid pieces and radiation and convection across voids and any attempt to determine conductivity by experiment leads to an "effective conductivity" of the aggregate. This quantity is usually handled like a true conductivity. It lies much closer to the conductivity of the gas phase present than to that of the solid but is higher the closer the packing. When the porosity is fine the part played by radiation becomes diminishingly small but in coarse aggregates at very high temperatures radiation probably plays an important part especially as the contact points between particles then become smaller in number.

14

Thermal Efficiency

THE attainment of high thermal efficiency is very desirable in all heating operations. This means that the calories available in the fuel must be directed as completely as possible into the stock with the least possible loss to the surroundings and after the heat has had the desired effect on the stock it should be re-used for some other purpose if at all possible.

Thermodynamic theory indicates that there is a limit to the efficiency with which heat can be made to do work and that this limit corresponds to thermodynamically "reversible" reactions. In this field a reversible transfer of heat would occur from one body to another if their temperatures were different by a vanishingly small amount—obviously not a very practical condition, as the transfer would be vanishingly slow and unless the system were thermally perfectly insulated from the outside world all the heat would leak away before it could be transferred in the desired direction. Under such conditions, however, thermal efficiency would be 100 per cent and there would be no loss of virtue.

It is worth remembering that the thermodynamic efficiency is best—or loss of virtue least—when the temperature difference between the donor and the acceptor is least. At the same time, the heat transfer rate is highest when that temperature difference is a maximum and, as heat loss to the surroundings is a function of time, there is obviously an advantage in working at a high transfer rate. Obviously, the best compromise solution should be sought.

One way of reducing the rate at which heat is lost to the surroundings is to use insulation or "padding" on the furnace, between the hot chamber and the outside wall. The use of a carefully chosen sequence of insulating refractories can reduce the heat flux to the furnace wall to negligible proportions but the effect on the temperature distribution in the refractories must be taken into consideration. The effect of insulation is that the high temperature isotherms open up and recede from the hot face so that the bricks there, which may have owed their survival to the fact that, say, seven eighths of their length was below their R.U.L. temperature may now be entirely above that temperature and suffer early collapse. To this the only answer is a better refractory, if there is one. In extreme cases there is none and it is only the temperature gradient imposed by the chilling effect of the surroundings which preserves the hot face brickwork. Open hearth basic roofs last longer if accumulating dust is regularly blown off them. Indeed it is not uncommon for air or water cooling to be used to impose an even steeper gradient for the preservation of refractories in really difficult conditions as found in the hearth and bosh of the iron blast furnace and in the doors and in the burner areas of the open hearth steel furnace. Neither of these would work satisfactorily without water cooling. The practice represents a serious attack on thermal efficiency, of course (even where the cooling system is operated as a waste heat boiler)—but fuel economy is never the *primary* purpose of furnace operation.

The use of insulation is well established where low or moderate temperatures are involved—in steam lines, hot blast mains and small batch heat-treatment furnaces, but it is also very appropriate and quite practicable in moderately high temperature units like recuperators, pottery kilns and some heat-treatment and billet heating furnaces, particularly where counter flow heat transfer is being applied (see p. 175). Then something approaching thermodynamically "reversible" conditions can be arranged and that the more efficiently if the slower rate of heat transfer can be exploited without heavy losses to the surroundings being incurred.

In furnaces like the open hearth steelmaking furnace the charge has to be heated to a very high temperature ($\sim$1600°C) and held there for several hours. Once the temperature has been attained heat is required only to meet losses through the furnace walls and roof. This is one case where insulation cannot be used very effectively for the reasons already given. Assuming, however, that the losses have been reduced as far as practicable, every calorie lost from the walls at say 150°C must be replaced inside the furnace by a calorie at 1600°C and the provision of this heat involves the loss of a very much larger number of calories in the flue gases. It is obvious that thermal efficiency can be best attained here by making the flame temperature and therefore the available heat above 1600°C as high as possible. As the flame temperature is increased so is the heat transfer rate and the proportion of the available heat which is actually transferred before the gas leaves the furnace. It is, of course, essential that combustion should be complete within the furnace and the gases should leave at a temperature not much above the bath temperature. The loss of virtue in this case is of little significance. The important thing is to supply as many calories as possible to the bath for each unit of fuel burned—and to keep the demand for total calories low.

Waste Heat

It is necessary, of course, that the heat in the gases leaving such a primary operation shall be put to some further use at as high a potential as possible. While this is customarily referred to as the utilization of "waste heat" it is, of course, only waste heat if it is not utilized. The principal applications of surplus heat are:

1. In recuperators or regenerators, especially for pre-heating the air (and/or gas) if high flame temperatures are required.
2. In "waste heat" boilers.
3. Miscellaneously in drying floors and for space heating.
4. For draughting.

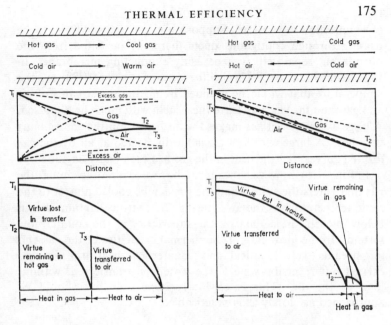

FIG. 21. Heat Exchangers.

(a) Parallel-flow. Final temperature cannot be higher than obtained
by mixing. Loss of virtue is very great.

(b) Counter-flow. Final temperature depends on relative heat
capacities of the two streams. If these are nearly equal loss of virtue
may be quite small, as heat is transferred at each stage without
much loss of temperature.

5. In some furnaces it is also used for pre-heating the solid
   charge as it approaches the hot zone and as the gases leave
   toward the chimney.

The transfer of heat from a body of hot gas to colder gas or to
water to make steam is carried out in equipment going under the
general name of "heat exchangers". There are several possible
designs of heat exchangers, but particularly parallel flow and
counterflow patterns.

A little thought and reference to Fig. 21 reveals the superiority of counterflow operation. Suppose hot gas and cold air to be passing through contiguous ducts thermally insulated from the outside but separated by a perfectly conducting membrane. In Fig. 21(a) where flow is parallel the air cannot be heated above the temperature that would be attained by mixing the streams of air and gas, and the gas may leave the system with much heat unused. A large excess of hot gas makes the air hotter but the gas remains very hot. A large excess of air is hardly heated at all by the gas (dotted curves) but the heat in the gas has been badly degraded.

In Fig. 21(b), however, the air can be heated almost as hot as the gas that enters the system and the gas can be cooled almost to the same temperature as the air entering. The virtue diagrams drawn below in each case emphasize the superiority of the counter flow system. In the ideal case, if the thermal capacities of the gas and air streams were identical and counterflow heat transfer was carried out reversibly, the virtue would be transferred, without loss, to the air. This is obviously impossible but should be approached as closely as is practicable.

Recuperators

Recuperators are counterflow air (or gas) pre-heaters in which the hot and cold gases flow in adjacent flues in a brick heat exchanger or through and around metallic tubes if the temperature is low enough. Brickwork tends to become leaky but acts as a reservoir of heat which minimizes fluctuations in the temperature of the heated gas. Metallic recuperators, offering low thermal resistance between the hot and cold gases, are theoretically more efficient. They are more compact and sound in construction (up to their limiting temperature which is about 700°C even using heat resisting steels) but they do give a degree of pre-heat which fluctuates in phase with the waste gas temperature. Design should be in accordance with the principles outlined in Chapter 13 to get high velocity turbulent flow on both sides of the tube's wall. The greater the effective area and the longer the time of contact on both sides and the more complete the insulation from the surroundings the more complete and efficient the transfer will be.

Regenerators

Regenerators are used at higher temperatures, beyond 1000°C pre-heat. The principle is that a mass of chequer work—a honeycomb pattern of brickwork is heated up by passing hot waste gases down through it for a fixed time. Cold air is then passed through in the opposite direction and is thereby heated up. The chequers at the end of the heating cycle are very hot at the top (~1500°C in open hearth chequers), but much cooler at the hot gas exit flue. Air entering is "warmed" by the first bricks it meets and gradually heats up till it leaves with a temperature which might be as high as 1200°C. Two chequer chambers work together, one on air and one on gas, the period between reversals being about 30 minutes. It will be obvious that this system is essentially a two-stage counter flow heat exchanger. Its efficiency will be subject to similar conditions as that of the recuperator, but there will always be some degradation of heat occurring during the thermal cycle as heat flows from the hot end to the colder end of the chequer all the time by conduction and radiation.

Regenerator design should aim at high velocity turbulent gas flow, the greatest practicable area of brickwork and maximum time of contact—or the best compromise among these. The bricks should have high heat capacity and adequate strength in their temperature range to carry the load above; resistance to slagging, especially at the top; and resistance to spalling (see Chapter 20).

Specially shaped bricks are available to meet these requirements but they are costly. Most chequers are built of bricks 5 cm thick which is the maximum thickness out of which heat can flow in half an hour (in both directions). Thicker bricks would run with a hot inactive core. The spacing is a compromise between close spacing for high velocity and wide spacing for low resistance to gas flow. High density bricks could enhance heat capacity but would put undue load on the bricks near the bottom. The over-all size and effective cross-sectional area should be adequate for the gas flow required through the chamber.

Cowper stoves are special forms of regenerators used in blast furnace plant to provide hot blast from the *potential* heat in the carbon monoxide in cleaned blast furnace top gas, rather than from sensible heat in waste gas of which there is none available. The chequer work is similar in design to that in steel plant regenerators but the shape of the chequer chambers, tall and narrow, probably gives rather more efficient transfer of heat (see Fig. 47).

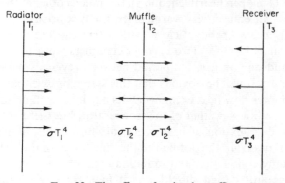

FIG. 22. The effect of a simple muffle.

Muffles

A muffle is a refractory screen placed between the source of heat and the stock in the furnace to protect the stock chemically from the flame and to give a more uniform distribution of temperature. The form would usually be that of a tube of round, rectangular or ⌂-shaped section inserted into the combustion chamber. In the simplified case (Fig. 22), the radiator, muffle and stock are presumed to be black bodies and the muffle to be a perfect conductor. If in the absence of the muffle the radiator and stock are at T_1 and T_3 respectively the net heat flow rate

$$H = \sigma(T_1^4 - T_3^4)$$

using the simplest formula. If the muffle is interposed it assumes an intermediate temperature T_2 such that the heat transfer rate is now

$$H_1 = \sigma(T_1^4 - T_2^4) = \sigma(T_2^4 - T_3^4)$$

$$= \frac{\sigma}{2}(T_1^4 - T_3^4)$$

$$= \tfrac{1}{2}H, \tag{14.1}$$

so that, under the ideal conditions described, the flow of heat is reduced by half and if n such muffles are interposed, $H_n = H/(n + 1)$. If a muffle has a thermal resistance and a temperature gradient develops across it the heat flowing from radiator to stock is even further reduced.

The above argument presumes that T_1 and T_3 are maintained constant. If the heat flow rate and T_3 were to be kept constant the temperature of the radiator would have to be increased, i.e. a hotter flame used or a larger radiation area. This would usually require more fuel and it is obvious that the cost of the benefits of muffle heating is high.

The same principle can be used to reduce heat losses from very high temperature sources. A few light screens or radiation baffles of refractory ceramic or metal arranged concentrically round the hot zone can reduce heat losses considerably.

Heat Balances

The assessment of fuel efficiency can be expressed in several ways. The simplest is the heat balance which is an arithmetic account of the various forms of energy entering and leaving the furnace. All energies must be expressed per unit of time or per unit of weight of products or in some other useful manner, and converted into the same units, whether calories or kilowatt hours, or coal ton equivalents. Energy supplies would normally be by electrical or chemical energy plus, if appropriate, pre-heat energy in the air or gas supply. There might also be chemical energy derived from reactions within the process. Energy leaving the system would normally appear as sensible and latent heat in the product and in the waste gases, and heat lost through furnace walls. Where there is any pre-heating or recuperation the proportion of waste heat circulated must be indicated. A full heat balance is useful in showing where economies may be effected.

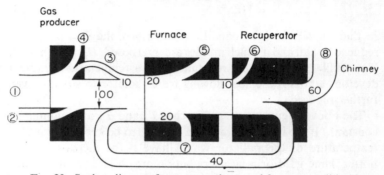

Fig. 23. Sankey diagram for a gas producer and furnace conditions already described in Fig. 10, curve (h).
(1) Potential energy in coal, (2) sensible heat in steam, (3) sensible heat in gas, (4), (5), (6) thermal losses, (7) heat to stock, (8) sensible heat to chimney gas. The loop shows 40 units of heat recycled in preheated air.
See also Fig. 24.

Sankey diagrams are diagrammatic heat balances. The rate of heat flow at any part of a process is represented on paper by a wide band—the width being proportional to the heat flow rate represented. Where different sources of heat are brought together —e.g. chemical heat in fuel and pre-heat in air—the two appropriate bands are drawn so as to merge together to a wider band of width equal to the sum of the components. Where heat is lost, a band of appropriate width is peeled off the main band and suitably labelled. Figure 23 shows a typical Sankey diagram for a gas producer, furnace and recuperator. The figures used are the same as were used in constructing Fig. 10 and also Fig. 24.

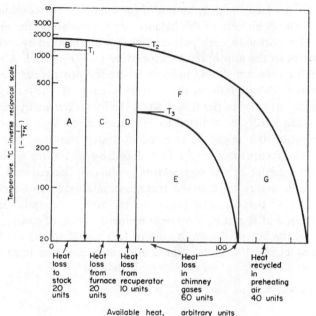

Available heat, arbitrary units

FIG. 24. The distribution of virtue.

In a hypothetical case, using the combustion gases of Fig. 10, curve
(h), and taking the potential energy in the fuel to be 100 units,
suppose the energy in the gases to be accounted for as indicated along
the heat axis. This is the same distribution as is represented in the
Sankey diagram (Fig. 15). The virtue is distributed as follows:
Loss on combustion is shown in Fig. 10. Virtue transferred to stock
heated to T_1 is given by area A and loss on transfer by B. Furnace
heat losses account for virtue given by area C and gases leave
furnace at T_2 with virtue D + E + F + G. D is lost by the recuperator; E is carried away by waste gases at T_3; G is recycled with
the heated air; F is lost in the transfer process.

Thring's virtue diagram (see p.114) can be constructed to provide
a complete description of the temperature, energy and virtue
distributions in a process. In Fig. 10, curve (h) defined the virtue
remaining in the combustion gases when the indicated fuel was
burned with pre-heated air. That curve is now transferred to

Fig. 23 in which the distribution of the energy of the combustion gases—the debit side of the balance—is set out along the energy axis. The virtue in each parcel of energy is obtained by erecting ordinates to the appropriate temperature $(-1/T)$ value. A study of such a diagram should indicate where the process can best be improved thermodynamically. In this case, if the energy is expressed as calories per ton of stock the loss of virtue by furnace losses can hardly be reduced by insulation at so high a working temperature but might be reduced by faster working. The loss D by the recuperator might be reduced by insulation while the loss on transfer in the recuperator could only be reduced by a change in design so that the transfer was effected more nearly reversibly so that T_3 came lower and the area G expanded left at the expense of E and F. Any improvement would, of course, save fuel and the potential heat in the fuel (per ton of stock) would be reduced below 100 units. Some authentic examples have been described by Thring* in considerable detail.

* THRING, M. W., *J. Inst. Fuel* **17**, p. 116 (1944).

15

Furnace Aerodynamics

SOME knowledge of the flow of hot gases in furnace systems is important to the furnace designer. The mathematical treatment is not fully developed but employs the methods of dimensionless criteria. The present discussion will be mainly qualitative.

Gas flow in furnaces is almost entirely turbulent but this turbulent flow may be steady or eddying. In steady flow there is a narrow layer of laminar (streamlined) flow along the boundary surface in which the velocity rises from zero at the boundary to the bulk velocity in a very short distance. At bends, or at changes in cross-section, unless these are very gentle, this laminar layer breaks away from the surface and an unstable condition ensues in which "eddys" break off the gas body and move into the "space" formed between the gas body and the surface. It is rather easier to imagine this sort of thing taking place in liquids where "cavitation" and its effects are easily demonstrated. The formation of eddys dissipates a lot of energy as heat and sound, and results in a severe loss of pressure.

The force driving gases through a furnace system is the algebraic sum of (1) the initial pressure, (2) the net pressures due to the various bouyancy effects where changes in level occur and including that in the final chimney, if any, (3) the pressures developed by any fans or injectors and (4) the pressure losses due to frictional resistance to flow. This might be written:

$$P = p_0 + \sum (\Delta p_b) + \sum (\Delta p_f) + \sum (\Delta p_r) \qquad (15.1)$$

183

where 0, b, f, and r denote initial, bouyancy, fan and resistance respectively.

p_0 can usually be ignored as it is normally the atmospheric pressure against which the others are measured but if only part of a system were under consideration it might have to be included in the calculation.

A contribution toward p_b is made at every section of the system which has a vertical component. The pressure is due to the difference between weight of the hot gas in the shaft and that of a column of cold air of the same height and cross sectional area outside. This is given by

$$\Delta p_b = 0{\cdot}95 p_a \left(1 - \frac{288}{T_g} \cdot \frac{\rho_g}{\rho_a}\right) h \qquad (15.2)$$

where ρ_g and ρ_a are the densities of gas and air respectively at S.T.P., $T_g°$K is the gas temperature and h is the height of the shaft. In the c.g.s. system:

$$\Delta p_b = 0{\cdot}00123[1 - (288/T_g)(\rho_g/\rho_a)]h \text{ g/cm}^2 \qquad (15.3)$$

$$= 0{\cdot}00123[1 - (288/T_g)(\rho_g/\rho_a)]h \text{ cm water gauge} \qquad (15.3a)$$

$$= 0{\cdot}00121[1 - (288/T_g)(\rho_g/\rho_a)]h \text{ m bar} \qquad (15.3b)$$

or if h is in feet

$$\Delta p_b = 0{\cdot}0147[1 - (288/T_g)(\rho_g/\rho_a)]h \text{ in. water gauge} \qquad (15.3c)$$

which is a commonly used unit. Other formulae may be used. Allowances may be made for variations in atmospheric temperature assumed above to be 15°C and for differences in atmospheric pressure from 760 mm/Hg. Where the bouyancy due to a high chimney is being calculated T_g is taken as the mean temperature of the gas in the chimney. Several ways of estimating this mean value have been suggested assuming in effect a temperature drop of between 0·5 and 1·2°C per metre height for brick chimneys, or about three times as much for steel ones. Obviously an estimate of the aeromotive force of a chimney cannot be a very exact quantity and will depend on the construction of the stack and to a considerable extent on weather conditions. Several shafts may be effective in one system including the chimney. For those in which the hot gas rises Δp_b is positive but where the gas passes downwards against the bouyancy the sign is negative.

Fans, turbo-blowers or air or steam injectors to accelerate the gases through a venturi throat may be used to supplement the natural draught or to replace it altogether where, for example, the waste gases have to be cooled and cleaned before discharge. The aeromotive force of the booster in pressure units is added to the net bouyancy to give the total positive force driving the gases through the system.

There is pressure drop due to friction at all points in a furnace system, least in long wide straight flues, greatest at sharp bends, sudden changes in section, junctions and baffles.

In straight sections the pressure loss is due to friction at the walls and is affected by the texture of their surfaces. At bends and baffles it is due to eddying in which a lot of kinetic energy is dissipated as heat and is then not available as either kinetic or potential energy when conditions steady up again.

In straight round sections the loss of pressure is given by:

$$-\Delta p_r = \rho_g \cdot V^2/2g \cdot L/D \cdot F \qquad (15.4)$$

where V is the gas velocity, ρ_g is its density, L and D are the length and diameter of the (circular) duct, and g is the acceleration due to gravity. If the duct is not circular D may be replaced both here and in the Reynolds number assessment by the "mean hydraulic depth" of the section, which would be $\frac{1}{2}(d.D/(d + D))$, D and d being the lengths of the sides in the case of a rectangular section. For simplicity the equation for the round section will be retained here. The value of the friction factor F depends on the Reynolds number and the roughness of the boundaries. This roughness is expressed as the ratio ε/D where the absolute roughness ε is the average height of projections on the surface. The ratio ε/D is called the relative roughness of the surface and is obviously of least importance in the widest ducts. The value of F has to be estimated from

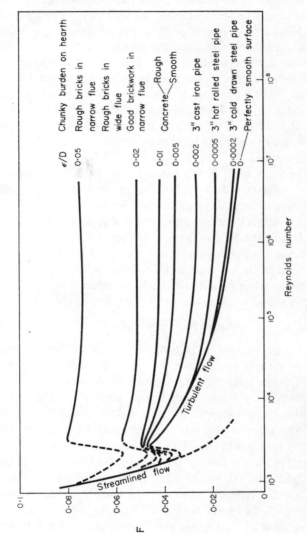

FIG. 25. Relationships between the friction factor F and the Reynolds number for surfaces in various states of roughness, indicated by the relative roughness ε/D.

an empirically determined graph of which versions are available in varying degrees of detail in standard reference books. The general form of this graph annotated to show how it is used is given in Fig. 25. The relationship between F and Re is independent of roughness as long as the flow is streamlined, F falling rapidly as Re increases to about 2000. When Re exceeds 2000 an intermediate condition extends up to about Re = 4000 beyond which, in the range of turbulent flow, there is a family of curves for different values of relative roughness. Even when the walls are perfectly smooth F is much greater in turbulent flow than if the streamlined flow relationship had been maintained. The value of F rises for ordinary "rough" surfaces by a factor of about 1·5 over that for polished surfaces and for very rough surfaces by a factor of up to about 2·5. When the Reynolds number is very high (high velocity gas flow or very narrow duct) F tends to level off especially when the surfaces are very rough, presumably due to the turbulence becoming complete or saturated. It will be appreciated that stock lying on the hearth of a furnace may constitute a very high degree of roughness, especially if irregular and "chunky".

At bends, junctions, or changes in section the formula for Δp reduces to

$$-\Delta p_r = \frac{\rho_g V^2}{2g} . k \qquad (15.5)$$

where k does not depend on Re (turbulence again being saturated in these locations). Values of k for a wide range of topographical features can be found in many reference books and textbooks,[4, 15] subject often to adjustment for geometric proportions, radii of curvature, roughness or edge effects. The loss of pressure at any of these places can be expressed as "k velocity heads" and the value of k varies from almost zero in cases where expansion or contraction is effected in funnels of very shallow taper ($< 6°$) called Venturi throats, through fractional values for well smoothed bends to values of the order of unity for sharp bends or sudden expansions. Sharp hairpin bends, the negotiation of a hole in a baffle plate, T-junctions and Z-bends give values of k ranging from 2 to 5. The values are higher on expansion than on contraction because it is the inherent difficulty in converting kinetic energy efficiently into potential energy which causes most of the eddying and energy loss. Smoothing of corners, rounding of edges and graduation of changes in section are all useful in reducing resistance to gas flow but these each take up space and are not always practicable to install.

Other formulae are available in the literature to assist in the calculation of the resistance of other features in furnace systems such as regenerator chambers, fuel beds and banks of boiler tubes. They need not be reproduced here. They are all of an empirical nature and should be used only under conditions similar to those from which they were originally developed. It will be appreciated that some of these features will normally have both bouyancy and resistance which are estimated separately.

The resistance of all the several features in a system can be added together to obtain its total resistance to gas flow. In so far as the separate parts are usually so close that the disturbance produced at one has not died down before the next one is reached, the simple addition of the partial resistances is most likely to yield a high value for the total resistance. In many cases an account of the pressure throughout the whole system will be required rather than an over-all balance. This must be obtained by working through the system from a point of known pressure taking into account as well as possible the effect of each separable item, adding or subtracting as appropriate.

Obviously the static pressure must fall through the system or the gas would not flow, but owing to the intermediate bouyancy effects and temperature changes which affect gas volume, the value of the pressure (which seldom exceeds a few inches water gauge) may fluctuate considerably and even assume negative values in places. Pressure changes are also caused by local acceleration and deceleration of gas flow brought about by change in cross section. If the cross section of a flue decreases the dynamic pressure head $\frac{1}{2}\rho v_2$ will increase and the static head fall by a similar amount. A sudden expansion due to a rise in temperature would probably cause an increase in both static and dynamic pressures. If the chamber were suitably shaped a large part of the kinetic energy could be converted to potential energy and the static head increased (as in the Venturi hearth furnace—Fig. 34).

It should be noted that in the above examination of the pressure distribution in a furnace it has been assumed that the velocity of the gases is known at each point, that is, the volume flow rate is known, and it is implied that this is steady at any point and that in fact a steady state exists. In such a case the acceleration of the mass of gas as a whole being zero we must expect P as calculated by eqn. (15.1) also to be zero at least within the error of the computation, otherwise the flow rate would either increase or decrease. The same methods could be used to assess the probable flow rate in a system if P were determined for a range of assumed flow rates until the rate which would give $P = 0$ could be interpolated.

The first essential in furnace design is that heat shall be developed at a required rate. This necessitates the combustion of a certain amount of fuel and so the design must be such that the fuel and air can reach the combustion chamber and the products of reaction escape from it at an appropriate rate and without unduly high pressures being necessary. Resistance to flow should therefore be as small as practicable except where some advantage is to be gained as in a regenerator, from some degree of constriction to promote turbulence and assist heat transfer. The gas velocity should not be excessive as that increases resistance, so flues must be adequately wide. High velocity also causes damage to brickwork through erosion by transported dust. Furnaces may operate under pressure (from a fan) or under draught (from a chimney). The aim should be that, either way, the pressure difference from atmospheric is as small as possible to minimize either drawing in cold air (which affects the temperature distribution, virtue of the heat energy, and overall economy) or escape of flame and gas through doors and brickwork (which damages the structure and causes a direct loss of energy).

In choosing dimensions of various sections of the system the temperature of the gas must be considered, and hence its volume. The furnace chamber itself is a zone of very sudden expansion while flues toward the chimney could become progressively narrower as the gas cools and contracts.

In any deep section of the furnace pressure differentials may develop normal to the flow direction, due to bouyancy. Further, distinct stratification of hot and cold gas can also occur, and in spite of turbulent flow conditions these may be persistent. The obvious case is where there is cold stock on the hearth. This cools the gases that come low over the hearth and these, having become relatively dense, fail to mix with the hot light gas in the roof. This is wasteful as available heat is denied access to the stock, and furnace design should attempt to minimize such an effect for example by making the hearth wide and the roof low or by deflecting the burned or burning gases downwards on to the hearth. The judicious use of supplementary jets can sometimes be effective in redistributing furnace gases.

The shape and direction of streams of gas issuing from burners have been studied theoretically and watched visually or photographically. Ideally a combustion chamber should be tailored to fit this stream (or vice versa). In practice it usually induces a good deal of eddying in its wake. The momentum of such a body of gas is considerable and for a jet of cold gas issuing from a nozzle of diameter d with an initial velocity V_0, the velocity at a distance x ($> 8d$) is given by:

$$V_x = \frac{8 \cdot 4 d V_0}{(x + d)}$$

At about $8d$ the velocity is undiminished but at $16d$ it is halved. If combustion occurs the gases must expand and gain in velocity approximately proportionately to the cube root of the volume increase, i.e. 2 to 3 times. A stream of cold gas would expand as a narrow cone with an included angle of about 9° but burning gas would give a cone about twice as wide. Observation of burning oil streams in open hearth furnaces has shown narrow, almost cylindrical streams of unburned gas penetrating far into the furnace, burning on the outside. The jet of burning fuel is deflected by the bouyancy and by the flow of gas already in the furnace. A hot jet into a cold furnace rises sharply to the roof but into a hot furnace rises more gently. The jet is also readily deflected by stock on the hearth or by cold air entering at doors and these can cause undesirable temperature distributions in the furnace chamber.

Rigid theoretical treatment is unlikely to provide an exact description of this process. In coming to a qualitative understanding of the behaviour of gas streams in furnaces great progress has been made through the use of models.[15] Scaling down is not as easy as it appears at first sight, however, particularly where elements of time and temperature are involved. The linear dimensions of a furnace may be reduced by a factor of $1/n$ but if the velocity too is reduced by the same factor the time of traverse is unchanged from the full scale to the reduced scale while the Reynolds number for the flow is $1/n^2$ times the original unless the furnace gas can be replaced by another fluid whose kinematic viscosity is $1/n^2$ times that of the gas. It is most important that Reynolds number should be simulated and at the same time it is rather an advantage if time can be extended so that observations can be made without resort to very special techniques such as high speed photography. There is little that can be done about the effects of temperature variations. Models usually work at room temperature and deductions about the further effects of high temperatures and temperature variations have to be made in other ways. The scaling down of the mass or density of a gas could be effected by using reduced pressure but not without reducing the kinematic viscosity. The scaling down of surface roughness by careful grinding of the appropriate surfaces in the model would be desirable.

Despite the difficulties, both air and water models of furnaces have been built and have yielded valuable information. Air models are easier to build. The mixing of air streams representing fuel and air can be "watched" by means of gas analyses of samples taken at selected points, for a tracer addition made to one of the streams, or by temperature measurements if the streams are initially at different temperatures. The effect of the sampling probe on the fluid flow pattern must be minimized by careful positioning. The side effects of combustion—temperature rise and sudden expansion—are not represented. Water models come nearer to dimensional similarity and can be operated with velocities low enough to be observed by eye or photographed by quite ordinary equipment using fine solid particles or air bubbles entrained in the water to render the flow pattern visible in a selected plane illuminated by a collimated sheet of light. Results of experiments both on furnace models and on simpler shapes of vessels show that gas flow in a chamber is seldom straight-through but that a large part of the gas circulates not only in the "dead" space in corners but in large volumes not occupied by the main stream. These circulating patterns must have important effects on temperature distribution and on refractory wear and the designer should aim to get the maximum benefit from them. The effects of bends, baffles and other similar features found in furnace constructions can also be examined in suitable models and some idea of what eddying looks like can be gained along with the possibility of modifying the design of these parts in such a way that the eddying is at least visually reduced in the model.

Results from the two types of model are in broad agreement and it has been possible to get some confirmation that the flow patterns at least in cold full scale plant are similar to those observed in the models.[15] The information obtained from a model is not quantitative but it does afford useful guidance to the designer and creates useful impressions in his mind as to what is likely to happen in the furnace he is about to build.

16

Furnace Construction

IT may appear from earlier sections that the ideal furnace is a
refractory box in which the maximum amount of fuel can be
burned as nearly as possible to the theoretical flame temperature
and the products of combustion removed with the lowest possible
residual heat content.

In fact, the metallurgical furnace has a job to do and getting the
stock in and out is as important as handling fuel and waste gas.
In many cases furnaces are machines in which materials are processed at high temperatures and in these cases compromise is
usually necessary between what is thermodynamically desirable
and what is possible under prevailing circumstances.

Furnaces usually have a cast iron or steel frame or case and a
refractory brick wall and lining. The choice of refractory is outside the scope of this section but should be adequate to the job in
hand and should be as good an insulator as it is practicable to use.
Thermal expansion in brickwork must be accommodated, usually
by leaving spaces between bricks during building otherwise the
frames will become distorted because the expansion of the refractories is likely to be three or four times the maximum elastic
strain in steel. In the case of arched roofs the shapes of the tapered
bricks should be so calculated that a good tight roof is obtained
at the working temperature and the distance between the skew-
backs must be adjusted to give a perfect arch with the correct rise.
This is done with adjustable tie-bars across the top of the furnace
between the vertical columns at the sides and ends (Fig. 32). In

193

other designs the rise of the roof is restricted by controlled downward pressure, while in a third type the bricks making the roof are suspended individually to give a roof of any contour, including flat.

Apart from the steel or cast iron skeleton of a furnace, metallic components are used in parts such as doors and door frames, grates, burners, skids, and mechanical rabbles and some of these have to work at quite high temperatures. When the temperature is very high, water cooled copper components are needed—blast furnace tuyères for example. At more moderate temperatures air cooled cast iron can be used as in most grates. Heat resisting cast irons are also available at a price which is undoubtedly worth paying for some purposes such as in sintering machine grates.

A furnace should have stable foundations appropriate to its size and weight and the main framework should be grouted into this and should carry the major part of the weight of the brickwork, as uniformly as possible. A sprung arch for example, would sit on the skewback channels fixed to the frame and *not* on the walls. The stack of a blast furnace sits in a mantel-ring which distributes the load through a ring of heavy columns, and not via the bosh wall, to the foundation. Small tilting furnaces rest on trunnions through which their weight is distributed to the foundations. Larger furnaces either rest on roller bearings or on rockers which would be simpler to maintain. In all cases the furnace case itself must be rigidly built using heavy steel plate, well supported and with the brickwork packed into it carefully to avoid the possibility of break-outs.

The use of steel sheet round brickwork minimizes gas leakage and allows low temperature insulation to be packed in if desirable. It can be painted with aluminium and can have good appearance but restricts access to brickwork.

Water cooled sections should be readily repairable in case of failure and precautions should be taken against cooling units becoming silted up. Cooling blocks should be of high purity copper where possible and the flow arranged so that the coldest water flows over the hottest surfaces with very high velocity.

Means for charging and discharging the furnace and if appropriate for arranging the flow of stock through it are usually straightforward engineering jobs. Special tools are used for lifting and laying ingots at soaking pits, carrying pig and scrap horizontally through open hearth furnace doors, and tipping it out of its box, or pushing billets through reheating furnaces; blast furnaces are charged with skips or buckets filled automatically and delivered through a system of bells designed to minimize segregation by sizes (Fig. 47). Suitable ladles must be available for receiving molten products and adequate crane capacity to handle the ladles and also service the furnace itself. The stock may pass through the furnace by gravity if the traverse is vertical or if sloping down an inclined plane (e.g. in rotating kilns) or it may be pushed along skids or carried on a moving hearth or on rollers if horizontal. Molten charges flow under gravity the hearth being so shaped as to be self-emptying once the taphole has been opened. Some furnaces tilt for pouring and occasionally liquids are discharged by syphon or suction arrangements (e.g. lead or aluminium).

Stirring or mixing the stock in roasting, or melting furnaces may be by rabbles or rakes (Wedge roaster) or by rotation of the furnace (rotating kiln, crucible and hearth furnaces). Molten stock is often mixed by reactions involving gas evolution. Electromagnetic stirring has been applied to large electric arc furnaces and of course, occurs in induction melting.

While refractories are expendable they are costly and it is essential that they should be well chosen and well laid with suitable refractory cements. The advantages from laying bricks well so that the roughness of the inner walls is low is apparent in the last chapter. Slagging and spalling may be inevitable but slag or spallings should not be allowed to accumulate to impede gas flow. The limitations of the bricks should be recognized and any special precautions necessary during firing or cooling rigidly observed.

The present day tendency is towards larger and more economic units, mechanically rather than manually served, and fully instrumented. The success of a furnace design is probably best measured (1) by the rate at which the furnace can consume fuel, (2) by the efficiency with which it transfers its energy to the stock, and (3) by the amount of fuel it can consume before it is worn out. Fuel costs must be related to capital charges as well as to costs of materials and labour. The efficient use of refractories can be as important as the efficient use of fuel and design should permit rapid repairs to be effected using "castables" or prefabricated sections where possible in order to minimize loss of production time.

17

Classification of Furnaces

FURNACES may be grouped in various ways, none of which forms a basis for a comprehensive classification system. They may be grouped by purpose (melting, roasting), by mode of firing (gas, coke, electricity), by shape (crucible, shaft), or by the method of charging (in-out, straight through, etc.). No matter how it is done, there is division and sub-division of each class, much overlapping and many miscellaneous misfits left over. The present discussion recognizes three major classes of metallurgical furnace, the crucible furnace, the hearth furnace, and the shaft furnace. Although these are distinguished mainly by shape even with this very simple classification it will be seen that there are furnaces which could be included in more than one group. To these must be added the minor classes, retorts distinguished by their volatile products; converters which are high temperature reaction vessels in which gas is blown through liquid metal; sintering machines in which air is drawn through an incandescent bed of mixed ore and fuel; and a small residue of types which will not fit into any of these classes. Finally, probably the most important class of all, there is the boiler.

Brief descriptions of various types of furnace in each of these classes are given and their modes of operation considered in the light of the theory discussed in earlier chapters.

Class 1—Crucible Furnaces

Crucible furnaces are used in foundries for melting small batches of ferrous and non-ferrous metals. The charge is melted in a refractory (often plumbago for its high conductivity) or metal pot which may be fired externally by coal, coke, gas or oil, or internally by gas or oil burners. Alternatively, the crucible may be integral with an induction furnace. Small units are fixed and the pot must be removed for pouring. Larger units are tilting furnaces from which the pot need not be removed and therefore suffers less mechanical and thermal damage.

Crucible furnaces vary from primitive pots in natural draught coke fires (Fig. 26) through a whole range of improvements, each involving more control over combustion (Figs. 27, 28). Distribution of heat before melting is mainly by natural convection within the crucible and is usually inefficient but in some modern

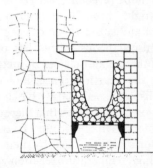

Fig. 26. Simple, natural draught, coke fired, pit type crucible furnace.

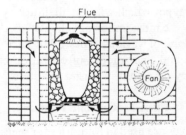

Fig. 27. Forced draught permits some preheating of the air and better control over combustion rate.

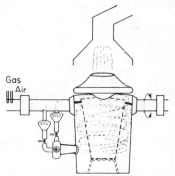

Gas
Air

FIG. 28. The use of gas or oil firing allows the tilting crucible furnace to be developed. Larger crucibles can be used and they last longer.

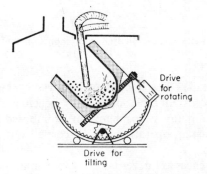

Drive
for
rotating

Drive for
tilting

FIG. 29. Rotating crucible fired internally has much better heat transfer, but the chemical effect of the flame could be objectionable.

furnaces the crucible rotates on an inclined axis while a gas or oil flame either swirls round the outside or plays directly on the charge (Fig. 29). These are more efficient, and useful for rapid melting of swarf scrap and even in some easy reduction processes. Induction melting furnaces (Figs. 30, 31) are, of course, in this class and are the historical replacement of the old multiple crucible furnace—a regeneratively fired chamber—in which tool steel was

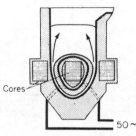

FIG. 30. Cored type normal frequency induction melting furnace
(see Fig. 15(a)).

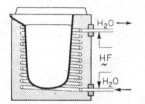

FIG. 31. Coreless type high frequency induction melting furnace
(see Fig. 15(b)).

traditionally made. The stock may, if necessary, be isolated from
the furnace gases in most designs of crucible furnace. With
induction heating any atmosphere can be imposed and the whole
melting and casting operation can be carried out in vacuo on a
moderate scale.

Crucible furnaces are never large—a maximum of 200 lb charge
for fixed furnaces and about five tons for tilting furnaces (except
induction furnaces which may go much bigger—see p. 147).

Class 2—Hearth Furnaces

This is the widest class of all and includes a wide range of fixed,
tilting, and rotating furnaces and kilns fired with all fuels and used
for roasting, melting, reheating and many other purposes. In
most cases the charge, on the floor or hearth of the furnace chamber
is heated by convection and radiation from the flame (or electrical

source) above it. In a minority of cases the fuel is mingled with
the stock and is burned by air passing over the hearth or blown
down on to it as in the case of the traditional blacksmith's hearth.

Hearth furnaces evolved from primitive concepts. The simple
"air furnace" (Fig. 32) is the prototype for the complex modern

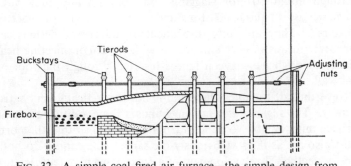

FIG. 32. A simple coal fired air furnace—the simple design from
which other hearth furnaces have developed.

giants. Furnaces of this shape were used in the "Old Welsh"
copper smelting process, bigger hearths being used for roasting
furnaces, smaller hearths for melting. The same type is still used
for melting copper alloys, but the firebox is usually replaced by
gas or oil burners (Fig. 33). Further development leads to the use
of regenerators for preheating the gas and air with heat from the
waste gases, and probably reaches its most complex form in the

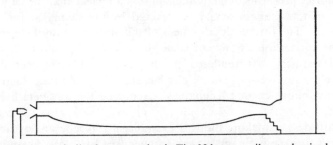

FIG. 33. A similar furnace to that in Fig. 32 but gas, oil, or pulverized
fuel could be used.

Venturi type regenerative open hearth steelmaking furnace in which both gas and air are preheated (Fig. 34). The traditional glass tank is of a similar design. Modern tendencies are towards the use of rich gas or oil or both together, only the air being pre-heated. These large steelmaking furnaces may be fixed or tilting (through about 30° for slagging and pouring) and may have capacities up to about 400 tons and even 1000 tons for inactive "mixers". They are fully instrumented and recent designs incorporate means of enriching the air with oxygen or injecting heat energy through gas–oxygen lances inserted through the roof or end wall.

Nevertheless, the over-all design is not theoretically very satisfactory. The regenerative system imposes the condition that the furnace must be able to be fired from either end so that the ports must also act as flues. Obviously the same ducts cannot do both jobs perfectly. The development of a practicable high temperature recuperator would make unidirectional firing possible. It would probably also lead to other changes. Firing across the furnace instead of along it should give more uniform temperature distribution with short hot flames. Some preheating of the solid charge might also become possible (prior to the use of the hot gases in the recuperators) and the process could become continuous. Thring* has proposed elegant designs incorporating these ideas but they have not yet been accepted for trial. There are some refractories problems outstanding and it is not clear that the capital cost of the changes would be covered by the saving in fuel at present. The further development of this furnace is unlikely ever to be carried out, however, since the march of progress in the steel industry has rendered it obsolescent. The regenerative principle survives in soaking pits but even there the introduction of continuous casting techniques is reducing the numbers being built.

Outside steelmaking the best example of a regeneratively fired hearth furnace is the glass tank furnace (see Fig. 35). These are

* THRING, M. W., *J. Inst. Fuel* **23**, 235, p. 406 (1960).

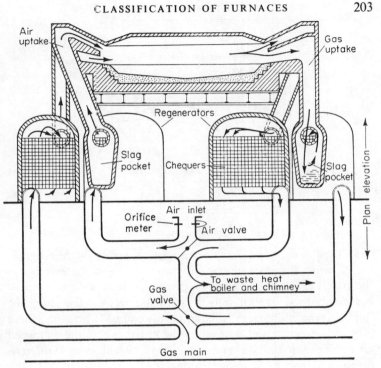

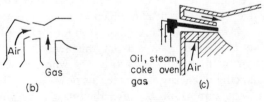

(a)

(b)

(c)

FIG. 34. (a) Schematic regenerative reverberatory open hearth steelmaking furnace, with "venturi" ports giving controlled expansion of burning gas.
(b) Alternative Maertz port, easily demountable for quick repairs.
(c) Alternative arrangement of port for firing with oil or coke oven gas.

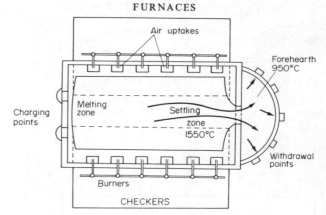

FIG. 35. Schematic glass tank furnace.

cross fired in the sense that the flames pass across the rectangular
hearth while the glass charge progresses along it. They have in
the past been fired with producer gas (preheated) but are now
commonly fired with oil, only the air being heated. The charge of
sand and metallic oxides is fed in at one end of the hearth where
they melt together and flow along the hearth where the entrained
air bubbles separate out and homogeneity is achieved. Different
firing rates can be employed along the furnace to provide the
appropriate temperature distribution. The glass at the holding
end may be raised to 1500°C. From there it flows under a skim-
ming bridge into a fore-hearth which may be separately fired to
maintain a lower temperature of only about 900°C. The glass is
withdrawn at various points from this section into moulding
machines or for manual fabrication. The whole process involves
a nice balancing of energy and materials flow rates. The choice of
refractories is important because the glass must not become
contaminated from that source.

   At the same time smaller reverberatories have been developed
for melting at the lower temperatures needed by, for example,
alloy cast irons and copper-base alloys. There are many types of
barrel-shaped furnaces (Fig. 36) fired with gas, oil or pulverized

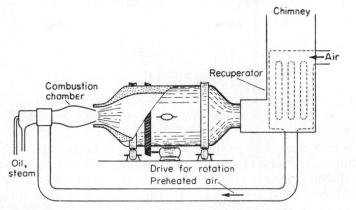

FIG. 36. Rotating barrel-shaped furnaces can be fired with gas, oil, or pulverized fuel. Air is usually preheated and if two bodies are connected in series the second can be used to preheat stock, and moved to the firing position when the first charge is completed.

fuel burned with recuperator preheated air. These are usually tilting, rocking or rotating types in which the unmelted stock can be moved to aid heat transfer and when molten the metal is made to wash round the lining. This permits very high heat transfer rates by convection without endangering the brickwork.

Billet heating furnaces have essentially the same form but the solid stock is placed on a flat hearth and either removed through the only door when heated, or pushed through the furnace from one end to the other on skids in the time required for heating— usually counter to the gas flow, the burners being situated near the discharge end. Some furnaces in which the burden passes through continuously are built in two or three zones as indicated in Fig. 41. The first zone might preheat the charge to about say 900°C in a billet heating furnace, the second bring it up to 1100°C and the third be designed to hold it at that temperature. All three burners would discharge to a common flue and the counter-flow principle would be observed. The height of the roof would have to be such as would accommodate the different volumes of gas

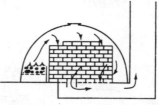

F<small>IG</small>. 37. Simple beehive kiln used in brickmaking and coking (obsolescent).

in each zone. Flat products may be progressed through such a furnace by means of a walking beam hearth. The work is periodically lifted from a grid on which it rests and advanced a short distance by a secondary support system whose movement is controlled by cams or eccentrics. This device allows some of the heating gases to be supplied beneath the work so increasing the heating rate.

Many kilns are specialized hearth furnaces. Even the beehive kiln (Fig. 37) has the same characteristics as the air furnace but its shape has developed to accommodate a large pile of brick or coal (for coking) and more convective heat transfer was accommodated. The beehive itself developed from the isolated batch furnace into battery kilns (Fig. 38) in which the air for combustion is pre-

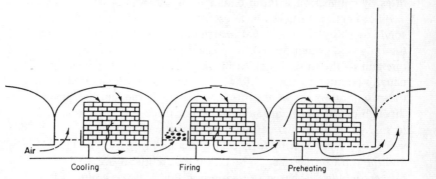

Air

Cooling Firing Preheating

F<small>IG</small>. 38. Development of beehive kiln. Only part of a battery is shown. These may be arranged in straight lines or in circles. The fire is advanced through the battery.

heated by using it to cool the previous charges while the waste
gases go to preheat the following charges. Finally this system
evolved into the tunnel kiln (Fig. 39) in which the stock is loaded
on to bogeys and passed through the waste gases, the firing section
and then against the cold secondary combustion air—the stock
moving through the fire instead of the fire progressing through the
stock. Tunnel kilns are used for drying and firing bricks and also
in one process for reducing iron ore. They are thermodynamically
very efficient counter flow heat exchangers.

Figure 40 shows how a tunnel kiln design can be curled round
into a circle. The stock would be passed through either on pallets
or on a moving hearth. It might be discharged cold as indicated
in the drawing but the design is more often used for heating billets
for forging in which case discharge is from the highest temperature zone and the heat in the gases must be recovered in some
other way.

Rotating kilns (Fig. 42) also follow the same pattern. They are

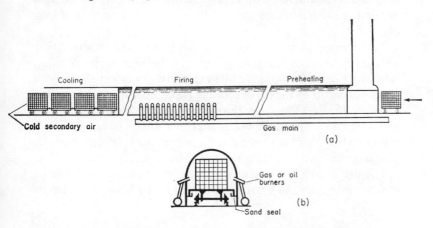

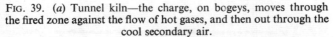

FIG. 39. (*a*) Tunnel kiln—the charge, on bogeys, moves through
the fired zone against the flow of hot gases, and then out through the
cool secondary air.

(*b*) Shows a cross-section in the firing zone. A more complex semi-
muffle design might often be desirable.

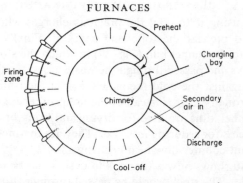

FIG. 40. Circular moving hearth furnaces operate in a similar manner to tunnel kilns. The firing zone may extend to the discharge point as would be necessary in, for example, a billet heating furnace.

long narrow cylinders—200 ft × 10 ft for example—rotating on a slightly inclined axis, so that the charge is transported with the aid of gravity counter to the flow of the burning fuel. Such kilns are used in cement and refractory manufacture and also in one method of making sinter and one method for the direct reduction of iron ore.

Two special forms of hearth furnace are used in extraction metallurgy. One is the multiple hearth "Wedge" type roasting furnace (Fig. 43) in which sulphide ore is transported across a series of hearths by means of rotating rakes, being passed from one hearth to the next lower one through holes alternately at the centre and periphery. The ore flows counter to a stream of hot air which reacts with its sulphur content and removes it as sulphur dioxide. The sulphur is the fuel and the process is called pyritic roasting.

The ore hearth is a smelting furnace in which ore and coke or coal are mixed on the hearth and heated by air blast which reacts

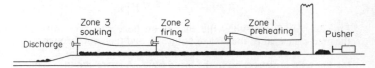

FIG. 41. Three-zone billet heating furnace.

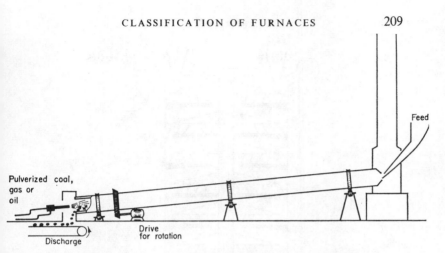

FIG. 42. Rotating kiln. In this long barrel-shaped kiln transport is by gravity. Stock moves against hot gases. Burners may be introduced in the middle of the length so that the similarity to a tunnel kiln becomes even greater.

with some of the coke, forming carbon monoxide which creates sufficiently reducing conditions to permit reduction of the ore. This is used for lead reduction, the special Neumann hearth being fitted with elaborate rabbling equipment to ensure complete reactions. The traditional blacksmith's hearth is essentially the same arrangement, the job being buried in a bed of coke burning in a stream of air injected through a tuyère.

There are several kinds of electric hearth furnaces of which the most important is the 3-phase Heroult furnace (Fig. 44) in which the 3-phase current passes between three large carbon electrodes through the slag and via the metallic part of the burden forming three large arcs of plasma with temperatures of several thousands degrees Celsius. This is the standard form used in steelmaking in sizes from a few tons up to about 200 tons capacity. This type of furnace was reserved for the production of high alloy steel until about 1960 except where electricity was particularly cheap. In that application there were some metallurgical advantages over

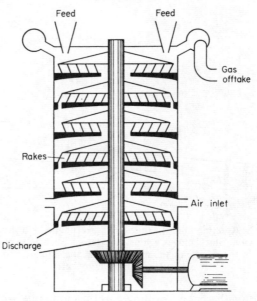

FIG. 43. Multiple hearth Wedge type furnace for roasting sulphide ores pyritically. This combines features of hearth and stack furnaces. The rakes turn over the charge and transport it through the furnace against the flow of hot gas.

alternatives, associated with the reducing conditions prevailing within which afforded some protection against loss of valuable alloying elements like chromium by oxidation. The cost of electricity made it unattractive for tonnage steel production but by 1960 it became apparent that relative costs of electrical energy, labour and capital had changed, and that the high price of electricity could be more than offset by savings in overheads gained by the faster melting rate under the electric arc and by the lower labour costs involved in charging operations which are simply a single drop from a basket while the roof is drawn aside. This applies, however, only to steelmaking from an all-scrap charge as the economies mentioned cannot be realized when

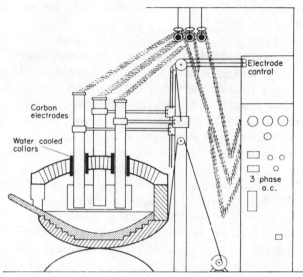

Fig. 44. Hall–Heroult type three phase direct arc melting furnace as used for steelmaking. Electrode positions are continuously adjusted. Charging may be via doors or through a removable roof. Furnace tilts for slagging and casting.

steel is made from a molten iron charge. There are also metallurgical advantages with respect to the quality which can be achieved under the reducing conditions available. The power consumption is high. A 100-ton capacity furnace would require a supply at the rate of 40,000 KVA at voltage tappings between 100 and 500 depending on the state of the charge. The current is controlled by automatic movement of the electrodes to positions which give the optimum arc length and hence arc resistance. During melting there are inevitably large surges of current which must be damped out by inductive reactors in order to safeguard the supply system. A major difficulty encountered in running the larger sizes of these furnaces is that of attaining a uniform distribution of temperature and composition in the bath. Electromagnetic stirring of the bath and rotating hearths are used to this end.

A second type is single phase with the arc struck between two carbon electrodes (Fig. 46) the energy being radiated to the stock in the hearth directly and by reflection off the roof. This is a clear case for rotating the furnace after melt-down in order to improve heat transfer and safeguard refractories. A third design also operates on single phase, the current being passed through the charge from a single electrode into a conducting hearth.

A fourth type of arc furnace has recently been developed for the matte smelting of copper concentrates (Fig. 45). In this the electrodes are effectively buried in finely ground calcine which is charged through the roof. The electrodes reach down to the slag level and current is passed through the slag and matte and back through the slag again. In this design no visible radiation escapes the furnace and process gases escaping through end flues are at a temperature of only about 200°C. The electrical arrangements may differ between plants. Primary supply would normally be 3-phase a.c. but in one design each phase is rectified separately and three pairs of electrodes operate separately (but simultaneously) providing d.c. arcs. Where costs are favourable this

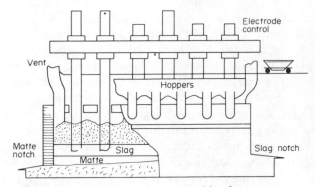

FIG. 45. Electric arc copper smelting furnace.

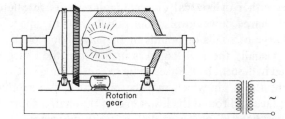

FIG. 46. Single phase indirect arc melting furnace similar in form to that in Fig. 36. Rotation of the barrel is essential after melting to safeguard the refractories.

type of furnace has replaced traditional oil-fired reverberatory matte smelting furnaces.

Heat transfer in arc furnaces is primarily by radiation from the arc except with respect to that part of the heat produced by the current as it passes through the stock. In the matte smelting furnace heat transfer is mainly by radiation which is almost totally absorbed. In the open hearth furnaces which burn fuel the heat transfer is also primarily by radiation provided the flame is luminous. This is supplemented by convection if the flame is suitably directed on to the stock. Radiation is reflected off the roof which should be able to withstand temperatures in excess of the maximum stock temperature. As long as the stock is cold the net heat transfer is from the roof downwards but when the charge melts its absorptive power falls and the temperature of the roof may rise dangerously because the bath is behaving like a mirror, reflecting all the heat of the flame back up to the roof. In steelmaking at temperatures near to the limit of the refractories it is necessary to keep the bath "on the boil" with bubbles of carbon monoxide continually bursting on the surface and creating as they do so transient but frequently recurring and quite effective black body conditions (see p. 166).

Class 3—Shaft Furnaces

These are used for melting, smelting and calcining or roasting.

They are either cylindrical or rectangular shafts through which the solid charge flows down under gravity into the space vacated by discharge of solids or liquids or by gasification of part of the charge. Usually the fuel (coke) is included in the charge and is preheated in descent by the ascending stream of hot gas ($CO + N_2$). The combustion air which may be preheated, is introduced via tuyères placed all round the lower end of the shaft a short distance above the "hearth" or "crucible", in melting furnaces.

The iron blast furnace (Fig. 47) is the most highly developed shaft furnace and can operate very efficiently not only as a heating unit but as a high temperature reaction chamber. Modern furnaces are nearly 30 m high (plus a considerable superstructure) and may be over 15 m in diameter at the hearth. Hot blast temperatures up to 1000°C are used and blast pressures may be as high as 3 bar with positive pressure throughout in some cases—the so-called high top pressure furnaces. The largest furnaces now smelt up to 10,000 tons of iron in a day requiring air at a rate up to 20,000 m^3/min. Recent developments include part-firing with oil or natural gas introduced at the tuyères and adjustment of air blast composition by adding oxygen and/or steam. These steps are designed to give the highest practicable combustion temperatures combined with good temperature distribution in the stack and afford more responsive control over energy input than is available through the coke charging rate alone. The charging system is important in determining the distribution of material sizes and hence the permeability of the burden to the combustion gases. Resistance to gas flow should be low and uniform across the furnace. The refractories are water cooled to give them greater resistance to wear, in the upper reaches, and to slagging, particularly in the bosh. The crucible wall also is protected against breakouts by coolers not shown in Fig. 47 and a circular palisade of "stave" coolers safeguards the hearth. Cooling water losses amount to about 2 per cent of the total throughout. Blast furnaces are tapped for slag and iron at regular intervals of about 4 hrs. Furnace campaigns last for several years, during which something like 2 million tons of iron may be made.

Heat transfer in this counter-flow process is potentially very efficient but the efficiency does depend upon the available heat in the ascending gases "matching" the demand for heat by the stock

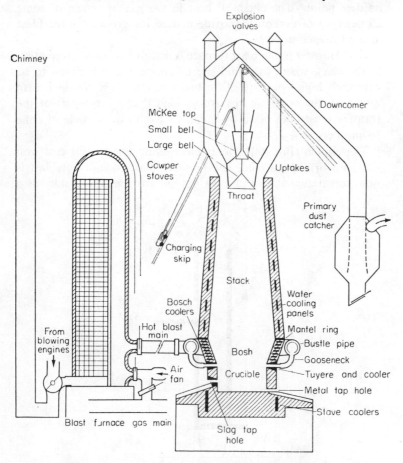

Fig. 47. Iron blast furnace with Cowper stove for preheating air, McKee type top for distributing the charge, and primary dust catcher.

at all levels. This can be controlled to a limited extent by those factors which determine combustion zone temperatures. The top gas temperature is about 200°C. It cannot be allowed to fall much lower because the gas must come past the dust catcher above its dew point. The chemical heat in the gas by virtue of some 25 per cent of carbon monoxide is used for preheating the blast in the Cowper stove.

Heat transfer in the blast furnace is mainly by forced convection in the stack where the wind speeds are generally believed to be extremely high—perhaps of the order of 150 km/h. In the hearth and bosh the temperatures are so high that a large part of the transfer must take place by radiation from one particle of coke to another.

The cupola (Fig. 48) is a simple melting furnace for cast iron usually working on cold, low pressure blast though hot blast is now sometimes used and even oxygen enrichment, usually as a

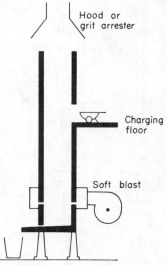

FIG. 48. Simple cupola for melting cast iron. Hot blast, auxiliary tuyères at higher levels and oxygen enrichment may be used to enhance efficiency and increase available temperatures.

control on temperature. Pig iron, scrap and coke are charged at the top over a large coke charge for preheating the furnace and providing a continuous reducing bed through which the iron passes to the hearth. These furnaces run for a few hours only at a time. On continuous operation their thermal efficiency would be much improved, but their design would have to be elaborated.

There are many designs of simple kilns (e.g. Gjer's, Fig. 49) based on the same principle, for burning limestone or calcining ore. In these cases the solid product must be withdrawn from the hearth mechanically. These kilns may operate on natural draught once started up.

Low shaft furnaces have evolved from high shaft furnaces already described, for use where a high degree of preheat and pre-reduction of ore is not necessary or not possible. One case is the experimental low shaft iron smelting furnace at Liège (Belgium) which was to produce iron from briquettes of ore and coal (which could not survive passage through a high furnace). A second case is the electric low shaft iron smelting furnace used in Switzerland,* Italy and Norway using cheap hydro-electric power to supply the energy for the process and coke only as a reducing agent. At this point the distinction between a hearth furnace and a shaft furnace becomes historical. Ferro alloys are produced in a similar type of furnace. In each of these cases a very large demand for high temperature heat is met by a high temperature source with low capacity so that its temperature falls very rapidly and there is no heat available for adequate preheating of the stock. Thermal efficiency suffers because the burden is not preheated and designs have been put forward for performing this preheat in a rotating kiln fired with the rich gas which comes off the furnace.

One other important shaft furnace is the gas producer, which is still used in steelmaking plants and glass-works, for example, though producer gas firing is rapidly giving way to oil firing.

Class 4—Retorts

In the production of gas from coal and in a few extraction processes

* GEHRIG, E., *J. Iron and Steel Inst.* **156**, p. 293 (1947).

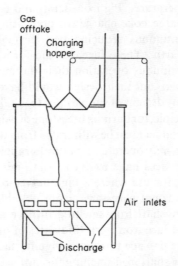

Gas
offtake

Charging
hopper

Air inlets

Discharge

FIG. 49. Gjer's kiln—a stack furnace for roasting ores—simple and efficient but largely replaced by equipment offering automatic discharge and a greater degree of control, e.g. sintering machines.

the reactions leading to a gaseous product are carried out in a retort. The traditional retort was a narrow vessel open only at one end, and the process was always a batch process. Retorts were operated in batteries with a common firing system. To a large extent these have disappeared in favour of a more continuous system in which the charge is passed down through a narrow chamber in which the reactions take place and from which the volatile product can be drawn without contamination by air. This applies in the coal gas and zinc industries but not so far in the coke industry which retains batch processing. The most recent developments in zinc production combine the blast furnace and the retort, lead being tapped from the hearth while zinc is collected from the top gas.

Class 5—Converters

Converters (Figs. 50-53) are used in steelmaking, in copper production and to a limited extent in other non-ferrous processes. Air is blown through or over the top of the molten charge, reacting with part of it to produce sufficient heat to maintain the process. In steelmaking the heat is derived from the oxidation of the elements silicon, manganese, carbon and phosphorus which it is desired to eliminate and the final temperature is several hundred degrees higher than the original. In copper smelting the elements oxidized are iron and sulphur. Heat transfer is

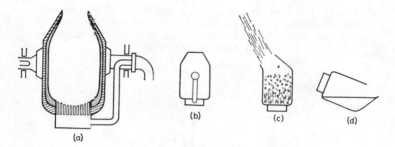

Fig. 50. (c) Simple Bessemer type converter for making acid steel; (b) Side view, concentric pattern; (c) Side view, eccentric pattern; (d) Side view, eccentric pattern, down for slagging.

excellent as it is produced within the charge but the losses in the waste gases are vast. Utilization of this waste heat for steam raising is now being adopted in some of the most recently built plant, but as it is only intermittently available elaborate provision has to be made to supply heat to the boilers from conventional fuel while the converters are not blowing. It may appear that these processes must be very economical on fuel but this is only partly true as carbon must be used in the blast furnace to reduce the silicon, manganese and phosphorus into the iron, and compressed air and oxygen also cost energy and money. The heat

220

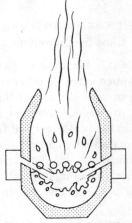

FIG. 51. Side blown converter used in steel foundries. CO burns in the vessel and gives good high temperatures. Capacities only 5–10 tons. A side blown converter is also used in copper extraction for burning out sulphur.

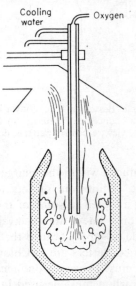

Cooling water

Oxygen

FIG. 52. Top blown (L.D.) converter—a recent development using oxygen, in steelmaking. Reduced refractory wear results from avoidance of bottom tuyères.

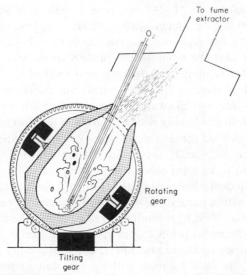

FIG. 53. Kaldo rotating top blown converter. Good heat transfer and better lining life (cf. the rotating crucible furnace, Fig. 29).

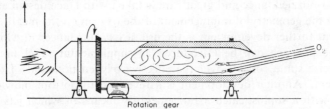

FIG. 54. Rotor converter combines the rotating hearth (Fig. 42) with oxygen top blowing.

available for the conversion of copper matte is however virtually without cost, beyond blowing costs.

Particularly in the bottom blown converters the wear on refractories is very severe. A new bottom may have to be fitted after every twenty or thirty heats and a complete lining with every fourth or fifth bottom. In this case there is nothing that can be done with water cooling and the best bricks are just not good enough.

The top blown or "L–D" converter is now used increasingly for steelmaking by the so-called "basic oxygen" or "BOS" process. This has replaced not only the bottom blown bessemer process but also the open hearth process to the extent that all new plant for refining blast furnace iron to steel is now of this type. The first stage in this development was the recognition that bottom blowing with air was introducing too much nitrogen into the steel. Bottom blowing with pure oxygen was too hard on the refractories around the tuyeres. The use of O_2–CO_2 and O_2–H_2O mixtures was technically successful but top blowing by oxygen alone proved to be most economical. The process has now been developed to deal with most qualities of iron and vessels are now being built with a capacity of 400 tons which can be processed in under an hour. This high production rate reduces the capital cost per ton to a figure far below that in open hearth steel making. Fuel and refractories costs are also lower, the steel quality is not impaired and the heat balance, without nitrogen in the reaction gases, is such that a high proportion of scrap can be incorporated into the charge. The oxygen is delivered through a water cooled, multi-nozzled lance and great care is taken with lance design and with the geometry of impingement of the oxygen on the metal. A recent further development is the use of oxy-fuel lances in which some propane can be introduced as supplementary fuel or to exercise some control over the effective oxygen potential of the system. Another development is a reversion to bottom blowing but with a sophisticated arrangement by means of which jets of oxygen are introduced through tuyeres within an annular shroud of an inert gas which delays the formation of iron oxide until the oxygen is well away from the refractory bottom.

Class 6—Sintering Machines

Blast roasting or sintering is carried out on certain ores either to improve their chemical composition with respect to carbonate or sulphide or to modify their physical condition. Ore and coke are charged on to a grate and ignited. Air is drawn down through the bed, burning out the coke, heating the ore, reducing it with carbon monoxide and finally re-oxidizing it after all the carbon is gone. This is carried out on a sintering "pan" or, more often, on a Dwight-Lloyd sintering machine (Fig. 55) in which the process is continuous. Heat transfer conditions are good, but contact time is short and thermal losses in the waste gas and in the sinter itself are very great.

Class 7—Some Miscellaneous Furnaces

(a) *Suspension Roasting* is applied to the conversion of sulphides to oxides or sulphates when the mineral has been very finely ground in the concentration process. The sulphide particles pass down through a heated chamber against a current of air, burning as they fall and providing the heat necessary for the continuance of the process.

This type of process suggests fluidized bed techniques in which gases and fine powders react while the latter are being carried along supported in the turbulent stream of the former. Heat transfer in these beds is extremely rapid and temperature control can be very accurate. This can be very important when chemical reactions are involved in the process. The technique is well established in chemical engineering and petroleum technology (where catalysts are entrained with the reactants) and a very considerable extension of its use in extraction metallurgy in the next few years seems likely. The limitation is that fine particles of many minerals tend to sinter together at quite moderate operating temperatures and the critical particle size distribution necessary for fluidization cannot always be maintained.

(b) *Soaking Pits* have developed from simple pits in the ground in which steel ingots were allowed to cool slowly, smothered in ashes. Modern soaking pits are regeneratively gas fired chambers each holding a number of ingots and still charged and discharged through the top. Heating is mainly by convective transfer from

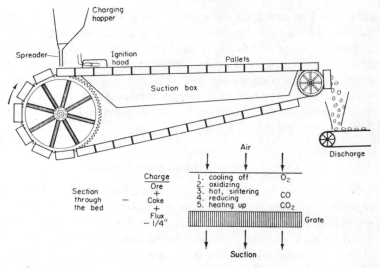

FIG. 55. Dwight–Lloyd sintering machine. Coke–ore mixtures are spread on pallets and ignited under down-draught. An extreme example of a "furnace" in which the chemical and physical processes take precedence over thermal efficiency.

the hot gases whose flow pattern must be such as to give the most uniform heating possible.

(c) *Forced air circulation furnaces* (Fig. 56) are rather like soaking pits in so far as many of them are charged and discharged through the top and heating is by convection. A powerful fan keeps the air circulating first over the electric heating elements, and then through the burden, which should be loosely packed so that the hot air can be well distributed over its surface for the greatest effect. The assembly of nichrome heating coils should also be arranged so that they lose heat to the air stream at the greatest possible rate. These furnaces do not normally go beyond about 700°C but the elements must of course run much higher than that. The efficiency of a furnace of this type is determined largely by its insulation.

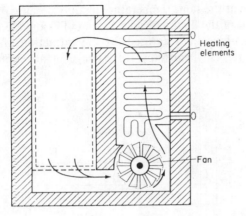

Heating
elements

Fan

FIG. 56. Pit type air circulation furnace.

(d) *Salt baths* are used for heating machined metallic parts out of contact with air which might affect the surface properties. They are commonly used in the heat-treatment of steels. The salts used are commonly mixtures of nitrates, chlorides, or sometimes cyanides. Nitrates cannot be used above about 500°C as they become unstable. Chlorides become volatile beyond about 800°C and cannot be used above about 900°C, except barium chloride which is used for heat treating high speed tool steels at 1350°C. Cyanides are used as a component to effect case-carburization. Glasses or slags might be used at still higher temperatures but are often rather viscous, and more fluid compositions are usually rather reactive toward either metal or containing vessels.

These baths may be in the form of simple crucible furnaces fired with gas or oil, the salt being contained permanently in the crucible which would usually be an alloy cast iron pot. Alternatively the salt may be heated by passing an electric current through it between two electrodes.

Heat transfer to the stock placed into the salt is by natural convection. This is very rapid—much faster than from a gaseous atmosphere at the same temperature, mainly because of the higher specific heat of the fluid (consider Eqns. (13.7) and (13.8)).

(e) *Electrolytic Cells* as used for the extraction of aluminium (Fig. 57) are also heated by the passage of an electric current

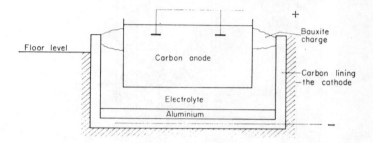

FIG. 57. Electrolysis cell for extraction of aluminium. The current used for electrolysis keeps the cell at 900°C.

through a molten solution of salt. In this case alumina is dissolved in cryolite and the bath operates at 900°C on the current used for electrolysis. There is of course no heat transfer problem and efficiency depends on the insulation provided and on other design features of the cell.

Class 8—Boiler Plant

Economically, steam raising is one of the most important uses to which heat is put. Technologically it is often one of the most inefficient of processes although large modern plant leaves little to be desired in this respect. A furnace for boiling water is usually called a boiler and this term covers everything from a simple water-jacketted firebox to the complex giants of the modern power stations.

There are three types of boiler. Small units for purposes such as central heating of moderately sized buildings are quite simple "shell" boilers in which the combustion space above the grate and the flue to the chimney are surrounded by a water jacket in the top of which the wet steam is collected. These are cheap, easily operated and maintained but not very efficient. At the expense of a little complication of design the heat exchange area can readily be extended either by fitting water tubes across the firebox or flue, or by passing the combustion gases through a number of parallel tubes running through the water space. Here then are the beginnings of "water-tube" and "fire-tube" boilers respectively.

Fire-tube boilers are used when larger quantities of steam are required at moderately high pressures. The construction of these is also reasonably simple and a number of designs have been in vogue for many decades. The Lancashire boiler (Fig. 58) consists of two wide flues passing right through the length of a cylindrical boiler drum, back along the bottom of the drum and then in two side flues to the rear again and to the chimney. The Scotch Marine boiler and the Economic boiler had narrower fire tubes carrying the hot gases through the water filled drum and the locomotive boiler was similar in principle. The construction of the drums of these designs was not suitable for use at higher pressures than about 20 bar. Heat transfer was mainly by convection. Superheaters could be fitted into most of them.

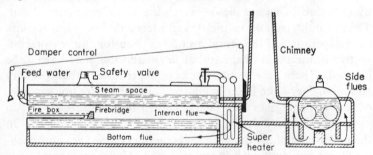

Fig. 58. Simple Lancashire, fire-tube boiler.

For higher pressure operation the water tube boiler is necessary primarily for reasons of strength. The boiler drums are aside from the path of the hot gases which are made to pass through one or more banks of tubes in which water is circulated by a pump from the drums. Another similar bank or coil of tubing constitutes the superheater in which wet steam is raised to its working temperature. The hot gas is also passed through an economizer in which feed water is preheated before entering the boiler drum, and finally through an air heater to preheat the combustion air. Heat transfer is almost entirely by convection. The heat transfer rate can be controlled on the water side by designing the tube diameter and water flow rate for any required Reynolds number. On the outside, high gas velocity would be attained by setting the tubes closer together. There is a limit to this however especially where there is the possibility of depositing ash or soot which would restrict the passage of the gases and impede the flow of heat.

The largest boilers used in central generating stations are water tube boilers with some differences (Fig. 59). The fuel is pulverized coal burned with preheated air to such a temperature that the ash is to a large extent slagged and collected at the bottom of the large vertical combustion chamber. This combustion chamber is lined entirely with long vertical water tubes so that its refractory walls are completely screened from the flame and its ash. The flame temperature being very high, heat transfer in this section is by radiation to these tubes in which the steaming is carried out. The gas must pass on to the superheaters sufficiently hot to raise the temperature of the steam over 500°C and for this reason convective heating of a final bank of steaming tubes may have to be omitted. There may be two banks of superheater tubes, primary and secondary—the latter being encountered by the hot gas first, of course. The gases then exchange their heat with feed water in an economizer and finally in the air heater in the usual way. In the most successfully designed units thermal efficiencies approaching 90 per cent are now being approached. Under these conditions fouling of the tube surfaces must be kept to a minimum in order to keep the over-all heat transfer coefficient high. If deposits are allowed to form on the water side hot spots develop and failure of the tube is imminent either by corrosion or by creep. Boiler waters are usually treated with chemical additions to minimize

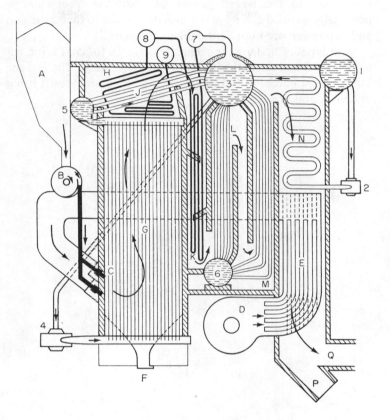

Fig. 59. Schematic diagram of large modern boiler. Coal from A is pulverized at B and burns at C with air from fan D, heated in section E. Gases pass up through radiation section G and through the system as indicated by arrows. Slagged ash is withdrawn at F. Dry ash collects at P. Gases at Q go to electrostatic precipitator and then to chimney. Treated water from (1) is pumped (2) through economizer N to main drum (3). It circulates in L, (6), M, and is pumped (4) through G, (5) and J—J, L and M being convection banks of boiler tubes. Steam is taken via (7) through super-heater K to (8); thence via secondary superheater J to live steam main (9).

corrosion. In the largest generating plant the feed water is practically distilled quality and is also de-aerated so that corrosion and deposition are both very unlikely to occur.

Most types of boiler were originally designed for coal firing but have been found adaptable to oil or pulverized fuel. The cyclone burner mentioned on p. 129 has been applied to steam raising in boilers of new design.[3]

18

Laboratory Furnaces

LABORATORY furnaces are of all types and are designed on the same general principles as industrial furnaces. They differ from these only in scale and in the degree of temperature control which may be necessary. Special requirements in respect of the control of atmosphere, the means of charging or discharging, and the siting of probes may have to be met, and in certain cases these conditions, coupled with a very high operating temperature may make design very difficult indeed.

Laboratory furnaces are usually electric as these are cleanest and easiest to control. Gas is used where flexibility is needed and especially where rather large spaces are to be heated to fairly high temperatures. Surprisingly good results can be obtained with gas or oil, but they create a lot of noise and do not admit much control of atmosphere unless completely muffled.

Most electric laboratory furnaces are resistance heated, using whatever resistor elements discussed in Chapter 12 are most appropriate. The more expensive materials like platinum, rhodium, tungsten, tantalum and silicon carbide are used for the highest temperature work. It is easier to find a resistor at 2000°C than to find a refractory on which to mount it. Indeed a refractory mount is best avoided if possible. At moderate and low temperatures the nichrome and kanthal type resistors are widely used.

A recent innovation for use at temperatures up to 1800°C is a composite element consisting of a molybdenum conductor enclosed in an impermeable alumina sheath. Electrical connections are made at one end of the sheath where there are also connections for hydrogen, a slow stream of which must be passed through the sheath while the element is hot. A number of these elements could be used together in a furnace. They would have to be protected from chemical attack (by almost everything) and from thermal and mechanical shock but this applies to anything that might be used at such high temperatures. The alumina sheath will be recognized as a muffle and to transfer heat through it to a furnace at 1800°C the molybdenum wire inside will have to be at a very much higher temperature but nevertheless the limitation is probably set by the refractoriness of the alumina rather than that of the molybdenum.

Induction furnaces are very useful for very high temperature work, being rapid in action, clean, and ideal for vacuum or controlled atmosphere operation. Temperature control to fine limits is, however, difficult. These furnaces are good for making small melts of steel and other alloys. Laboratory units are usually supplied by spark-gap or valve oscillator frequency converters, and use frequencies up to 100,000 Hz. The induced secondary current may be produced in the metallic charge or in a suitable screen of carbon or molybdenum (*in vacuo*) from which the heat can be transferred by radiation to the stock which then need not be electrically conducting.

Refractories in laboratory furnaces are again of all types but vitreous silica and aluminous porcelain meet the needs of a high proportion of cases. Beyond 1350°C refractory ware must be carefully chosen from sillimanite, mullite and alumina or special refractories like thoria and zirconia. These materials can be obtained as crucibles or tubes (with or without a closed end) which may be impermeable to gases if required. Impermeable tubes have poor resistance to thermal shock. It is obvious that resistance to slagging at very high temperatures is nil and furnaces must be kept very clean. Castable alumina cement (alundum), alumina mixed with a little clay as a bond, is commonly used for making refractory artifacts in the laboratory but these are not impermeable. The alundum is made plastic with water, moulded to shape, dried and fired.

The designing of a laboratory furnace requires some experience, not the least in guessing what power will be required to attain the desired temperature. The power input depends on the size of space to be heated and the working temperature and also on the degree to which the furnace can be insulated. Thermal efficiency is sometimes depressed by the necessity of having one end open for working or even occupied by probes which conduct heat away from the hot zone. As a guide it may be said that under favourable conditions a 25 mm diameter tube furnace at 1000°C would require about $\frac{1}{2}$ kW while a 75 mm diameter furnace at 1500°C would need 3 or 4 kW.

To achieve uniform temperature over a length of tube much greater than its diameter some care has to be taken—the furnace is usually wound differentially (closer toward the ends); radiation baffles are mounted along the axis; and a conducting insert used if practicable.

High temperature insulation using ordinary insulating materials is not very successful as most of these are either not very refractory or are weak at high temperatures. A series of radiation baffles or "muffles" will cut down the transmission of heat (see p. 178) and sets of concentric cylinders of molybdenum or platinum can be used mounted outside the source of heat. These have the advantage of low heat capacity and therefore do not lead to slow heating of the furnace. They are particularly appropriate under vacuum where there is no convection problem. A simpler high temperature insulator is fine powder of any refractory material—magnesia, alumina, carbon, etc. In effect, this constitutes a complex multiple radiation baffle, but it would also smother convection currents in a gaseous atmosphere. It is desirable in all cases that baffles should be mounted at the ends as well as round the sides. Convection losses within the working space should also be avoided, probably best by arranging that the hot chamber be fairly well filled by stock. Vertical tubes are particularly liable to losses by convection especially if the top has to be left open.

Control is discussed in Part 4, and the principles expounded there apply to laboratory equipment. Usually the aim is to operate off mains voltage if possible. Gauge and length of resistor would be chosen so that the wattage on full voltage was a little more than estimated necessary to attain the required temperature. Manual control could then be effected by series resistance, energy regulator or auto-transformer. In a well designed furnace there is little to choose between these provided it is working at design temperature. For flexibility, however, the series resistance is wasteful of energy and the energy regulator involves switching current on and off and may be hard on windings. Some form of automatic control operated by a thermometer in the furnace is usually desirable, preferably altering the energy input only fractionally. The efficiency of any such control system obviously depends on the thermal inertia of the system and this should be taken into account when the furnace is being built, and kept as low as practicable.

PART THREE

Refractories

19

Classification of Refractories

THERE is no general definition of a refractory. Essentially it is a material of "high melting point", but this is a relative term and melting point is not the only criterion of usefulness.

Most refractories are ceramic materials made from high-melting-point oxides, particularly SiO_2, Al_2O_3 and MgO. Carbon is now an important refractory, however, and carbides, borides and nitrides are also being developed for high-temperature work. Metals like Mo and W are refractory metals and find uses in research apparatus. Even these may be melted in heavily water-cooled copper containers which, if not classified as refractories, certainly replace them.

Finally there are materials like asbestos which have not a high melting point, but which are used as medium or low-temperature insulation and should not be excluded from consideration.

Refractories may be classified by chemical composition and in this book the broad division is into those based on silica, alumina and silica together, and magnesia and chromite either separately or together. This leaves a residue, much of which can be incorporated into the most usual classification which depends on behaviour toward metallurgical slags as follows:

Acid refractories are based on SiO_2 and include silica, and the fireclay series with 30–42 per cent Al_2O_3, sillimanite and andalusite with about 60 per cent Al_2O_3.

Basic refractories are based on MgO and include magnesite and dolomite, chrome magnesite and magnesite chrome. In addition alumina and mullite bricks are classed as basic and many "special" refractories like ThO_2 and BeO would come into the same category.

Acid refractories react readily with basic slags and basic refractories are attacked by acid slags as a general rule.

Neutral refractories would be relatively inert to both siliceous and limy slags and this class includes carbon, chromite ($FeO.Cr_2O_3$) and forsterite ($2MgO.SiO_2$) bricks.

"Special" refractories are usually new, or very expensive materials, such as the ZrO_2 and BeO mentioned above, and are reserved for research purposes, and other occasional uses such as in atomic energy and gas turbine technologies. These could also be classed as acid, basic or neutral, but this classification would not have any great practical significance. The term super-refractory is now being used for some of these materials whose melting points lie above about 1900°C.

For industrial use refractories are largely marketed as bricks of a variety of standard shapes and sizes or in non-standard sizes at somewhat higher cost. There are many standard shapes including tapered bricks to facilitate the construction of curved walls and arched roofs free of gaps between the bricks. Other standard shapes such as tubular sleeves for stopper rods have very specialized uses. Some materials are supplied in granular form often described as "pea sized" and this is built into a furnace while it is very hot by throwing thin layers on to the hearth and letting these frit into position. The same preparations, mixed with hot tar as a binder, are used to line the hearths of new furnaces. Some refractories are made up as "castables"—plastic preparations which can be made up to any shape and are dried and fired *in situ*, and are useful for effecting running repairs. For every kind of brick there should be a cement of similar chemical characteristics but preferably of plastic consistency. They may be made up suitably tempered with water and possibly other additions or may be supplied dry for preparation on the building site.

Some refractories are designed to have very low thermal conductivity. This is usually achieved by trapping a high proportion of air into the structure. Naturally occurring materials like asbestos are good insulators but are not particularly good refractories. Mineral wools are available which combine good insulating properties with good resistance to heat but these have no rigidity. Porous bricks are also made which are rigid to high temperatures with reasonably low thermal conductivity. These together are classed as insulating refractories.

Laboratory ware and research equipment are available in the forms of crucibles and tubes with other shapes to special order. These articles may be porous or impermeable and gas-tight. The new and the unusual, special both with respect to material and shape, are frequently in demand for aircraft engine, atomic energy and rocketry purposes. For these purposes, however, pure substances and mixtures of pure substances are required and the term refractory is not good enough so that we find the materials here called "ceramics", the meaning of which is gradually being widened to include practically any non-metallic inorganic substance of high melting point.

20

Properties and Testing

FAILING a definition of a refractory material, a discussion of the qualities sought in it is essential. The most important requirements of the user are:

1. Rigidity and maintenance of size, shape and strength at the operating temperature, which will presumably be "high".
2. An ability to withstand thermal shock such as is met in heating up and cooling down of furnaces, or in fluctuations which occur during charging or during normal operation.
3. Resistance to chemical attack by whatever gas, slag or metal is likely to be encountered.

As there is no supreme material capable of standing up to every possible condition, the choice must be made to meet the requirements of the job to be done. Compromise is usually necessary. Other properties such as cold crushing strength, and thermal conductivity are usually of secondary importance but occasionally, as in insulating bricks, one of these may become of prime importance. The properties actually measured are sometimes only an indirect guide to quality and must be interpreted with care. Thus high porosity will lead, *aliter parsim*, to reduced slag resistance and lower thermal conductivity, and this latter in turn to poorer spalling resistance. But higher porosity may be achieved by lower firing temperature resulting in less vitrification which in turn leads to high permeability (worsening slag resistance) but alters the mechanical strength to improve spalling resistance. Empirical tests have been developed—some of them not very satisfactory—for slagging, spalling, etc., which are intended to be a more direct guide to behaviour than fundamental measurements. The following tests are commonly carried out. B.S. Specifications are available for some of them marked ‡ and details can be found in B.S.S. 1902 (1952), or in an appendix to "Refractories—Production and Properties".* A similar series of tests is specified by the American Society for Testing Materials.

1. *Visual examination* should indicate general uniformity, vitrification, texture, etc. This is obviously a job for an experienced eye, but variations within a batch or between batches which are obvious are probably significant.

2. *Dimensional accuracy* is important when large structures are to be built. A ruler or calipers would be used. Edges should be straight and free of chips. Faces should be flat. Tolerances demanded can be very small indeed because, of course, the furnace has to be the correct shape and size when the requisite number of bricks are placed together. If the shape is good and the faces flat, bricks can be set dry, i.e. without using cement.

3. *After expansion* (or contraction)‡ is easily determined by heating a suitable sample of bricks for a prolonged time at the proposed working temperature. The bricks (or cut samples) are measured before and after treatment to determine the permanent change in dimensions which should, of course, usually be very small. The test must be related to working conditions. If the expansion (usually a contraction) is too great, more thorough firing is required, or another kind of brick. High values lead to severe cracking of furnace walls during use, or to distortion of the structure. A small expansion is preferred, to give tight joints, and for some special purposes like ladle linings a fairly high value is deliberately arranged.

4. *Reversible thermal expansion*‡ is determined versus fused silica (whose coefficient of expansion is very small) in any standard type of dilatometer. Here one is determining the coefficient of

* See bibliography 21.

thermal expansion along with any volume changes due to polymorphic transformations. The latter may give evidence of imperfect firing (e.g. in silica). The expansion is a necessary design datum as it must be accommodated as the furnace heats and cools during operation. It is also related to spalling resistance, and a low value is desirable on all accounts, but every type of brick has its own characteristic behaviour in this respect, and in some cases, e.g. magnesite, an otherwise admirable brick may be difficult to use because the coefficient of expansion is inherently high (see Fig. 60).

Refractoriness.‡ It is not generally practicable to measure the melting point of a refractory. Usually melting extends over a range of temperature—perhaps over several hundreds of centigrade degrees. The determination of solidus and liquidus temperatures would not be easy and might not be very useful. A material may be partly liquid yet appear solid. At a higher temperature where the proportion of liquid exceeds a critical value which may depend on its viscosity, the material will either collapse and appear liquid or pasty, or will exude liquid and collapse slowly later. The apparent temperature of collapse may well depend on the rate of heating, particularly if the liquid is very viscous.

The test for refractoriness is to compare the sagging of a "cone" of the material (either cut from the solid or made up from powder and a bond of dextrin) with that of standard Seger cones, when they are heated together, at a standard rate in an oxidizing atmosphere, until the test cone bends over. The number of the best matching Seger cone is quoted as the refractoriness of the brick; or its nominal melting point (which can be checked by pyrometer) is referred to as the Pyrometric Cone Equivalent (P.C.E.) Temperature. As will be seen later, this is usually much higher than the working temperature of the brick. Refractoriness is a guide to quality rather than a measure of usefulness. The refractoriness of basic bricks is always very high and seldom quoted because such differences as there are beyond 1750°C are of no practical significance.

Cold strength is measured by a simple compression test called the Cold Crushing Strength test. It is seldom needed to assure the user that the brick will not fail in compression but results may be used to show whether or not the brick has been properly fired.

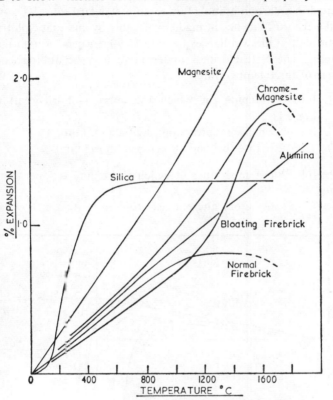

FIG. 60. Typical expansion curves for various refractories.

The test might also indicate whether the brick would transport readily without damage to corners and edges.

Hot strength is very important in high-quality bricks but is not measured directly. Instead it is a temperature that is determined— that at which deformation under a standard load is rapid. The test is called the Refractoriness under Load‡ test (or R.U.L.), and is rather like a short time creep test.

A constant load of 100 or 200 lb (or 25 or 50 lb/in.^{2}) is applied to a prism of brick 2 in. square by $3\frac{1}{2}$ in. high. The specimen is heated in a carbon granule furnace at a standard rate (10°C/min) and a record of its height made, preferably by automatic plotting, until the test piece collapses or sinks to 90 per cent of its original length. The Refractoriness under Load may be quoted as the range of three temperatures:

1. Initial softening (at which the curve is horizontal) (see Fig. 61).
2. Rapid collapse (contraction rate 0·05 in./min).
3. Total collapse (beyond 10 per cent shortening).

Sometimes the temperature at which shortening is 5 per cent is noted.

A more satisfactory alternative is to stop heating at a suitable

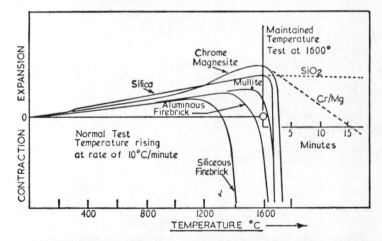

Fig. 61. Typical R.U.L. curves for various bricks. The behaviour of silica and chrome-magnesite under sustained load at 1600°C are compared in the inset.

temperature, related to the working temperature, and to measure the creep rate during the next few hours. The object is of course to determine the maximum temperature at which the brick could be used under compressive load. It will be appreciated that usually not all of a brick would be at the hot-face temperature. The "cold" end would normally remain rigid even after the hot end had exceeded initial softening. The extension of the period of these severe conditions from hours to weeks is not readily predicted and this leads to difficulty in correlating test results with practical performance.

Hot strength depends on the structure of the brick as well as on its melting characteristics and may persist beyond the solidus temperature if the major constituent has recrystallized during firing to form a three-dimensional network of interlocking crystals of high melting point which can maintain its rigidity even while a part of the material is molten in its interstices. The crystals of hexagonal tridymite and orthorhombic mullite behave in this desirable manner in acid refractories but the minerals in basic bricks usually have cubic habit and do not interlock, but, like glasses, deform slowly but continuously under load at high temperatures.

Other mechanical tests have been mentioned in the literature but are not in common use. Long-term creep testing would obviously be useful. The measurement of elastic constants has been reported as a check on thoroughness of firing. Abrasion tests using modified shot-blasting equipment have been described by Mackenzie* who successfully compared carbon and firebrick in favour of the former and observed that the most resistant part of the firebrick was its skin. Where bricks are to be subject to severe abrasion such a test is obviously worth trying though simple correlation with performance in the furnace may not always be obtained. Mackenzie noted a correlation between his test results and a transverse breaking stress determination which might afford a

* MACKENZIE, J., *Trans. Brit. Ceram. Soc.* **50**, p. 145 (1951).

better measure of abrasion resistance and strength generally, at elevated temperature.

Thermal shock resistance is measured by a *Spalling Test*.‡ There is no fully standardized test but materials can be compared using constant test conditions and these should be chosen to suit the brick under test and the working conditions under which it is to be used. For example, silica bricks have excellent thermal shock-resisting properties above 300°C if properly made, but shatter if cooled quickly below that. Any spalling test on silica must avoid cooling under 300°C. Similarly, a furnace would not be cooled below 300°C with silica in it. Various tests used involve heating either a brick (or several) or a prism cut from bricks, slowly up to a working temperature, and then cooling it rapidly for 10 min in a cold air jet, or on a steel plate, then reheating for 10 min in the furnace held at the working temperature. This is repeated until a piece of the brick becomes detached. The number of cycles to failure is noted. A brick surviving 30 cycles is claimed to be very good and quoted as having a spalling resistance of +30 cycles. The size of sample and severity of the heating and cooling must affect the result, and are usually much worse than met in practice. In some tests only one end of the brick is heated and cooled which again exaggerates the temperature gradient. A more elaborate test involves building a small wall of the bricks and heating it with gas burners, and subsequently cooling with air blast. This would normally be designed to simulate some particular practice. In this case heating and cooling are applied to the face or end only and not along the sides as well.

The result of a spalling test must depend on at least four properties of the brick—(1) its thermal conductivity which determines the temperature gradients set up under the test conditions; (2) the coefficient of thermal expansion which determines the strain induced by these thermal gradients; (3) the modulus of elasticity and (4) the shear strength of the brick substance, which together determine the stresses set up and whether or not they will be relieved by crack formation or propagation. The formation of cracks during early cycles would probably modify the temperature distribution during the later cycles in favour of even more severe stresses so that cracks, once formed, are likely to get worse. The mechanical properties of the material depend in a complex manner on the chemical nature of the substance and on its crystallographic condition as well as on the general distribution of grains of mineral, bonding matrix and pore space, or, in more practical terms, strength depends on porosity, and in thoroughness of firing, and where appropriate on the distribution of "grog" particles which help to arrest cracks.

Slag resistance is also difficult to test other than qualitatively. There are three types of test. The most common involves drilling holes in the brick and packing these with samples of typical slag, likely to be encountered. The assembly is then heated to a working temperature for about an hour, cooled and sectioned. The extent of the penetration of slag into the brick is noted and compared with others.

The second method is to mix ground brick and ground slag in various proportions and to determine the melting points or refractoriness of the mixtures. A graph of melting point vs. per cent slag can be drawn and graphs for different bricks—or different slags—compared. The choice of slag is of course appropriate to working conditions.

This second test is a measure of the inherent reactivity of the substance of the brick to the slag in question. The first takes some account of the effect of brick texture and slag surface tension as they affect reaction rates. It is still rather severe, however. It provides a breach in the normally dense skin of the brick, and it employs a brick in which there is no temperature gradient—the cooling effect back from the face would normally slow up slag reaction.

The third type of test again tries to simulate practical conditions and either individual bricks are dipped in slag, or hot slag is sprayed on to hot walls built into specially designed furnaces. Field trials may incorporate panels of test bricks into working equipment for comparison with standard types.

The extent to which a given brick will be attacked by any particular slag depends first on the chemical reactions likely to take place between them, secondly on whether any of the products of reaction are liquid at the working temperature and thirdly on the speed with which the reactions are likely to proceed. Generally acid slags react with basic bricks and basic slags with acid bricks but this is an over-simplification. Iron oxide, for example, one of the more corrosive reagents, may occur in either acid or basic slags and can damage both acid and basic bricks—or be contained by either type of brick in suitable circumstances.

The behaviour of bricks versus slag can in principle be predicted from the appropriate thermal equilibrium phase diagram. Figure 62, for example, might be used to deduce that iron oxide would rapidly form a slag with a silica brick, and this could be confirmed by employing a test using the second method mentioned above. It is a matter of experience, however, that silica

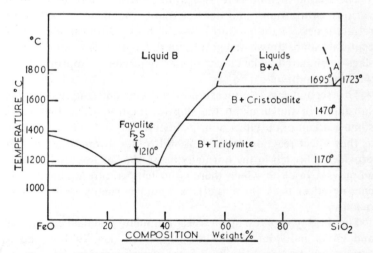

FIG. 62. The FeO–SiO$_2$ equilibrium diagram (after Bowen and Schairer).

bricks stand up remarkably well to iron oxide attack presumably because the slag which forms at the face of the brick is very viscous and reduces the rate of attack to a very low value.

In most cases a ternary or higher order phase diagram would be necessary from which to determine probable slagging behaviour and very careful interpretation is usually needed. Two examples are discussed in the section on firebricks for which the appropriate systems have been investigated and phase diagrams published, but in many cases the diagrams have not yet been worked out in sufficient detail, especially when we consider that most refractories contain quite a lot of "impurities" which are often potential fluxes. Even where adequate information is available in the form of phase diagrams their interpretation is not simple. Phase diagrams describe equilibrium conditions but slag attack does not involve much equilibrium. A cursory inspection of the liquidus surface or a few isothermal sections is seldom sufficient. Appropriate vertical sections must be drawn and an effort then made to deduce the phases present between the unchanged brick and the unchanged slag after some sort of dynamic equilibrium or steady state condition has been set up, and to estimate the proportion of solid to liquid and the probable rheological properties of this mixture of phases (see page 267).

True porosity‡ is a popular test on bricks and is generally considered to give a good index of quality. Its measurement involves the determination of real density and apparent density by standard methods.

Porosity is determined in the manufacturing stage by the size grading of the raw mineral, by the moulding pressure, and by the firing temperature and its duration, the effect on these last two being very much influenced by composition. Low porosity achieved by grading and moulding pressure is valuable in conferring high resistance to abrasion, to slagging, and to attack by gases such as occurs in the iron blast-furnace. Low porosity due to a high flux content and over-firing would not enhance the slag resistance but would probably give a very poor spalling resistance.

Porosity values are usually about 20–25 per cent with lower values down to 10 or 15 per cent obtainable with some difficulty when thought to be desirable.

Real density‡ is determined on crushed and ground material by the density bottle method. In firebricks the value obtained may reflect the degree of vitrification and in silica bricks it is used as a guide to completion of firing (see later), but generally the value is of little significance on its own.

Apparent density‡ of porous materials is difficult to measure accurately. Evacuation in a vacuum desiccator and flooding with paraffin to fill the unsealed pore space, followed by weighing in air and in paraffin, is a common method. The value obtained is of little use unless perhaps to calculate loads for transport. The same data can be used to calculate *apparent porosity* or, in conjunction with real density, *true porosity*, including closed-pore volume. Recent demands for low-porosity bricks have led to values down near 10 per cent being achieved with difficulty. High resistances to abrasion and corrosion are thus obtained, and impermeability toward gases, but the spalling resistance is severely impaired.

Permeability‡ is measured in a simple apparatus in which air is blown at a measured rate through a prism or cylinder of brick, the pressure drop being measured by a manometer. There seems to be little justification for the test except that, taken into conjunction with porosity, "texture" may be deduced—e.g. low porosity and high permeability may indicate cracks; high porosity and low permeability, closed pores or very small pores. Permeability may vary with direction through bricks made by the dry-press method, indicating stratification. Where reaction with gas is possible, low permeability is of course very desirable.

Thermal conductivity‡ is usually measured on whole bricks. The apparatus used comprises a hot plate on which the brick sits on its largest face, with insulation round about designed to ensure parallel flow of heat out through the brick at least at the centre. On top is a brass plate to distribute temperature evenly, and in the centre a blackened copper disc insulated thermally from the plate, but in the same plane. The hot-face temperature, and that in the disc can be measured with fine thermocouples. Draught screens are erected round about and the temperature of still air over the assembly measured. The heat loss from the disc by radiation and convection is obtained by empirical formulae and the thermal conductivity calculated using the heat through-put and temperature gradient after equilibrium conditions have been established. At high temperatures more elaborate design is necessary to ensure parallel heat flow through the central section of the brick.

The value of thermal conductivity obtained is an effective conductivity insofar as the mode of heat transfer through the typical refractory with its 25 per cent porosity is only partly by conduction through the mineral grains and across the bond from one grain to the next. At high temperatures convection in the pore space and particularly radiation across the interstices become important. There is therefore a relationship between conductivity as measured and the structure or texture of a brick. The inherent thermal conductivity of the brick substance is the overriding factor determining the normal value for any type of brick and in this respect basic bricks will be found to differ significantly from acid bricks. Bricks of very high porosity are made when low conductivity is required (see Table 13).

Chemical analysis‡ of refractories is carried out by standard classical methods, but spectrography or X-ray fluorescence techniques could be employed for some elements by manufacturers carrying out large numbers of analyses and special techniques like flame photometry for alkalies could be useful alternatives. Results of analyses indicate primarily the type of brick, e.g. the percentage alumina in a firebrick classifies it as siliceous, ordinary, aluminous, etc. Details of the analysis may also indicate the quality of the brick within its class, e.g. the level of alkalies and other potential fluxes has a strong effect on refractoriness of, say, aluminous firebricks. The overall analysis is not a complete description of quality, however, and especially in basic refractories the distribution of constituents is sometimes of much greater importance than their proportions.

Petrography. The petrographic microscope can be used on thin sections (one-thousandth of an inch) of brick to find the nature and distribution of minerals in the material. This can be applied either to the new brick or to follow changes occurring in firing or in service, or under attack by slags. Refractive index and isotropy are used to aid identification of minerals present.

Reflected light microscopy has also been used on polished sections but identification is dependent primarily on the petrographic technique.

X-ray diffraction can also be used to study mineralogical constitution and changes which may occur in firing or in service. The diffraction photograph would consist of series of "lines" for each mineral present above about 2 per cent and the system could be made semi-quantitative if necessary, but could not demonstrate spatial distribution.

21

Manufacture of Refractories

EACH kind of brick is made from a different raw material and its treatment usually involves some very special features, but there are so many common features that a consideration of the general principles will now be made.

Raw materials are mineral deposits—clays, sands, ores, rocks—which must be mined or quarried and then crushed. Many of these materials are imported, usually in bags to maintain purity, and some are very expensive. Fireclays are mined in this country and usually bricks are made at the mine head. Other types of brick are often made by the same companies in the same areas where coal is usually available and often the markets are handy also.

Crushing is carried out by simple ore-dressing equipment and material should be graded and stored by size.

The composition of the brick is adjusted by blending the materials along with any flux or bond additions that may be required. In many cases there is only one mineral used with appropriate bonds added. Blending is, however, not only by composition but also by size and the properties of the product depend very much on this stage. Low porosity can be attained by using a "close-packing grading" such that interstices between the largest particles are filled by a smaller grade, and residual interstices filled by a smaller grade again and so *ad infinitum*. Thus there is some control on the ultimate porosity—and on strength, for the maximum number of contacts will also be made with the closest packing. In basic bricks the coarse and fine fractions are

253

sometimes of different composition and the fine fraction ultimately acts as the bond. "Grog", or prefired materials, may also be mixed in at this stage. This may consist of broken and crushed scrap brick but sometimes specially hard fired material is prepared for the purpose. "Chamotte" has long been used in Europe. Recently "synthetic mullite" grog has been produced in Britain, fired at 1750°C to assist in the manufacture of high-duty refractories of the mullite type with high abrasion resistance and low permeability to gases.

Blending is carried out in a paddle mill with a kneading action where addition of water and bonds (sometimes temporary) are added. Clays attain some plasticity at this stage. Other mixes remain friable.

The next stage is moulding. Standard shapes are machine moulded, non-standard shapes by hand. Hand-moulding is most successful with plastic mixes usually rather wet (14–20 per cent water), which can be "thrown" into the wooden box-type mould and relied on to fill it. It is cheaper than machine-moulding on a jobbing basis, but the machine is preferred, as usual, for mass production. Semi-plastic machine-moulding with about 10–12 per cent water uses moderate moulding pressures and usually a two-stage process—extrusion to a rough shape followed by pressing to an exact shape. The dry-press method is used for all the non-plastic basic mixes as well as clays with not more than 5 per cent of water. Mixing in so little water is not easy and usually a fine spray or mist is used. When high pressures of the order of 350–1400 bar are applied in the mould, air is liable to be trapped and cause laminations. Because of this, clay may be de-aired by vacuum treatment either before moulding or, better, using vacuum blocks in the mould wall. Green strength is rather low in the bricks so that it is difficult to maintain good edges and corners, and handling of the green bricks should be mechanical as far as possible.

The only other important forming process is slip-casting, applicable mainly to clays which can be formed into a colloidal suspension with water and poured into a mould of plaster of

Paris which absorbs the water and causes a uniform deposit of clay to build up in the inside of the mould. This is obviously most useful for awkward shapes and for hollow ware. It is also used for special refractories which are prepared from very fine powders.

The bricks must then be dried. This is conducted either on large drying floors (heated by waste heat from kilns) where the bricks are laid out in open array, or in tunnel kilns where they are stacked on bogeys and passed through a tunnel against a stream of hot air. This is the faster and more compact process but cannot readily be adopted where sizes and shapes are not fairly constant.

The final stage is firing. The traditional furnace was the bee-hive kiln but this has been superseded by numerous developments. In one case there are a large number of kilns side by side. The firing kiln at any time is heated by a fire of coal or coke, using air preheated by being drawn through several cooling kilns. The combustion products are passed over bricks in several preheating kilns before passing to the stack. Heating must be uniform in each kiln so that the bricks attain a uniform temperature before the fire is advanced to the next compartment. (See Fig. 38.)

The other important type of kiln is the tunnel kiln with the firing zone of oil burners in the middle of its length. Cold secondary air is preheated as it cools the fired bricks on their way out. Combustion gases preheat the advancing cold bricks. The heating cycle and time at top temperature are very important and vary from one type of brick to another. The firing temperature should be at least as high as that at which the brick is to be used to ensure completion of reactions and attain as high a degree of dimensional stability as is practicable. (See Fig. 39.)

The cooling rate is critical in some cases, particularly for silica bricks which undergo phase changes at low temperatures. Cooled bricks are ready for use. For some purposes they are graded for size into, say, three ranges within the limits allowed by the appropriate specification. This is to facilitate the accurate construction of furnaces.

Increasing use is being made of refractories which are not manufactured into fully fired bricks. The chemically bonded basic brick is one example. In this case $MgSO_4$ is incorporated as a temporary bond which gives the green brick enough strength to enable it to be built into the furnace where during use the bond will decompose leaving MgO behind which reacts with other constituents to create the same kind of structure as is found in kiln fired bricks. These bricks may also incorporate steel reinforcing rods and steel casings which oxidize during use, the FeO formed becoming incorporated into the structure of the brick either to strengthen it or to bond it to adjacent bricks. A cheap variant of this is to pack steel tubes with magnesite and build furnace walls with these. On heating they oxidize and fuse into a serviceable monolithic structure. The L–D converter is now usually made from green, tar bonded blocks and ladles are now being lined with sand bricks made from suitably clay bonded sands—often natural sands can be found of suitable composition. The bricks may be built green into the ladle or, even more cheaply if practised on a large scale, the sand mixture may be "gunned" into position round a thin sheet steel former. The green sand must be dried and then "fired" by using the ladle. This seems to give satisfaction equal to conventionally brick lined ladles.

Refractory cements are also coming into more common usage. These were originally developed for patching and repairing. They have now reached such a stage of development that they are as good as that which they replace and it is sometimes convenient to use them when first building a furnace. An example is the throat of the blast furnace where the complete absence of jointing is probably an advantage. These cements are based on "ciment fondu" bonded mineral mixes; for example crushed "chamotte" or sillimanite would be mixed with calcium trisilicate $3CaO.SiO_2$, a "purified" cement. This mix can then be tempered with water, laid in position and allowed to set like concrete. On being heated the hydrated bond will decompose, leaving behind a fired refractory concrete the properties of which will be dominated by the type of mineral used.

Special refractories are made in the same general manner as has been described but the starting materials are very carefully purified "chemicals" rather than raw minerals. The size and grading are finer and more closely packed and the mixture must contain some kind of temporary bond which will be driven off during firing, leaving no harmful residue.

It seems likely that more extensive dressing of the minerals for ordinary refractories will be practised in the future so that closer control over composition will be achieved. This seems to be technically desirable for most basic refractories but the cost of such purification processes would be very high. These bricks are already very costly and any further increase in price would have to be adequately matched by improved performance in service.

22

Alumino-Silicate Refractories

The Alumina-silica System

The Al_2O_3–SiO_2 equilibrium diagram is shown in Fig. 63, with the composition of the various refractories marked on it. Up to 1545°C firebrick consists of mullite and silica. Mullite is present as needle-shaped crystals which form an interlocking matrix which can retain its strength and rigidity to high temperatures toward 1810°C at which it melts incongruently. The strength depends, of course, on the proportion of mullite present, i.e. on the Al_2O_3 content. Silica is present as tridymite or cristobalite (or both) (see page 274), but the effect of fluxes is to transform some of it to glassy silicates—thermodynamically very viscous liquids.

Above 1545°C liquid forms—again a very viscous liquid but viscosity falls off rapidly with temperature—and the structure becomes a matrix of mullite needles with liquid in the interstices. Considering siliceous firebrick (25 per cent Al_2O_3) we find that the proportion of liquid is about 70 per cent and it is unlikely that the remaining 30 per cent in the form of mullite will be able to maintain the shape of the brick for very long. The refractoriness of the brick has been exceeded at 1545°C and indeed probably at a lower temperature in the presence of fluxes. A brick containing 45 per cent Al_2O_3 on the other hand contains only 40 per cent liquid at 1545°C, the remaining 60 per cent being mullite, which remains solid. Temperature increases beyond 1545°C cause the amount of liquid to rise in all cases, but this rise is slow. At 1700°C the high-alumina brick has still only 45 per cent of

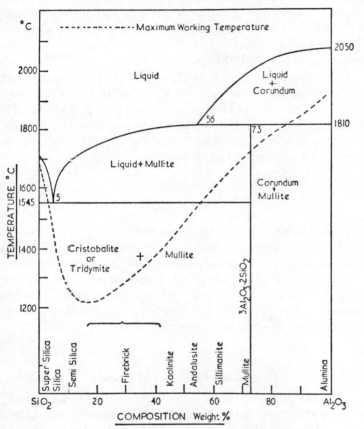

FIG. 63. The Al_2O_3–SiO_2 equilibrium diagram (after Bowen and Greig) indicating the compositions of various kinds of brick based on the system, and their approximate maximum working temperatures which depend also on other aspects of composition.

NOTE. A number of modifications of this diagram have been suggested in recent years, in particular with respect to the melting of mullite which has been reported as being congruent at 1850°C and with respect to the range of composition occupied by a mullite solid solution which would be expected from theoretical considerations. It has also been suggested that the eutectic should be at 10 per cent SiO_2 rather than 5 per cent. Since these proposed changes have not been accepted unanimously the original version is retained here and particularly as it is as satisfactory as any of the others as a basis for a discussion on the properties of alumino-silicate refractories.

liquid (although the other has 80 per cent and has undoubtedly slumped). It is clear that refractoriness in advance of 1700°C is quite possible with firebrick of suitable quality but R.U.L. values better than 1545°C will be difficult to attain.

Considering even higher Al_2O_3 content, at 72 per cent, at the mullite composition, the proportion of liquid at 1545°C has fallen practically to zero and remains negligible up to 1810°C. The advantages of very high alumina are obvious and the development of bricks with Al_2O_3 content beyond the firebrick range will be discussed at the end of this chapter.

Firebricks

Firebricks are the commonest class of refractory—the "common brick" of the furnace designer. They are made from fireclay which usually occurs in association with coal measures and is plentiful in Britain though quality varies considerable, the best deposits being in the central Scottish coalfields. China-clay deposits in Cornwall have also been used in recent years to make a similar type of brick with enhanced properties.

Origins and Properties of Fireclays

Clays are produced by the decomposition of igneous rocks by geological agencies and conditions not commonly encountered today. The kind of clay depends on the kind of parent rock and presumably on the treatment it has had. Fireclay is derived from acid rocks like granite, in which felspars of the type $K_2O.Al_2O_3.6SiO_2$ are decomposed by H_2O and CO_2 (possibly at high pressure and temperature) to give K_2CO_3, SiO_2nH_2O, and $Al_2O_3.SiO_2.2H_2O$. This last is the formula for kaolinite and also describes dickite and nacrite which have slightly different crystal forms, having been produced under different physical conditions. The structures of these clays are similar in that each is made up of alternate layers of "silica" and "gibbsite" ($Al(OH)_3$). The latter has itself a layered structure, a sheet of aluminium ions (Al^{3+}) being sandwiched between similar layers of hydroxyl ions (OH^-), each Al^{3+} ion being associated with 6(OH^-) ions, each of which has a share in two Al^{3+} ions. The silica layers are also made up of three sheets of ions. The middle sheet is of hexagonally arranged Si^{4+} ions with O^{2-} ions tetrahedrally arranged round them so that on one side there are three O^{2-} ions, each sharing two Si^{4-} ions, and on the other side one O^{2-} ion per Si^{4+} ion, this one unshared. The net effect is that there are too many O^{2-} ions for electrical stability and the silica sheet achieves its stability by incorporating its unshared O^{2-} ions into the gibbsite layer where they replace some of the (OH^-) ions. There are several geometric arrangements whereby this can be achieved and these give rise to the slightly different minerals named and to halloysite which has rather more water incorporated in its structure. There are other groups of clay with different chemical composition, particularly the Montmorillonite-Beidelite group which also occurs in fireclay. These have formulae as follows:

Montmorillonite $Al_2O_3.5SiO_2$ nH_2O (CaMg) O.
Beidelite Al_2O_3 $3SiO_2$ $4H_2O$.

The structure of these minerals is again layered "silica" and "gibbsite" but the order is different, there being more silica to accommodate so that there are two silica layers for every gibbsite layer. Further, aluminium ions may be partly replaced by magnesium ions which leaves an excess negative charge to be balanced by sodium, calcium and other cations which become loosely incorporated in the structure. These differences make this group of clays more plastic than the kaolinite group and their presence in fireclays renders them more workable than, say, china clay, but their lower alumina content and higher proportions of alkalies and alkaline earths impair refractoriness.

The clay material may have been purified and concentrated by the leaching out of soluble matter. The resultant "residual" deposit contains silica, micas, and clay, and is low in iron, alkalies and alkaline earths. The granular silica can be separated by elutriation as in the china-clay pits and the clay obtained is highly refractory, fires white and is mainly used for china. Otherwise the decomposed rock may have been washed away and separation of its constituents effected by differential settling in rivers or lakes. Such "transported" clay deposits are free of granular silica but may be contaminated by other substances, picked up during transportation, which impair refractoriness. The montmorillonite clays particularly can incorporate basic ions into their structure. Iron is a very common contaminant whose soluble salts are rather easily hydrolysed, depositing ferric hydroxide, often in small nodules.

Apart from essential clay substances many accessory minerals may be present in great or small amounts. These include free SiO_2 and Al_2O_3 from further weathering, iron oxide, carbonate, sulphide, titanate, and sulphate and salts of alkalies and alkaline earths, free TiO_2 and carbonaceous matter up to 15 per cent. These affect plasticity, refractoriness, and burning colour, and determine the usefulness of a deposit.

The very pure, white-firing china clay has been mentioned. The least pure clays derived from more basic rocks or heavily contaminated are red-firing and go to building bricks and drainpipes. Intermediate grades are the very plastic ball clay, bond clay, flint fireclay, ordinary fireclay and aluminous fireclay, all with commercial uses.

Plasticity in clay corresponds to fluidity in a liquid. In liquids, flow occurs under all positive stress, velocity V being proportional to applied force P. Hence $V = K\phi P$ where ϕ is fluidity (Fig. 64(a)). In clays $V = K\mu (P - f)$ where μ is mobility and f is a yield stress below which there is no flow (Fig. 64(b)).

Both μ and f depend on the water/clay ratio of the mix (64(c)), but different clays behave quite differently (64(d)). Plasticity also depends on pH (64(f)) and grog additions increase the effective water/clay ratio and act like additions of water (64(e)) (grog being

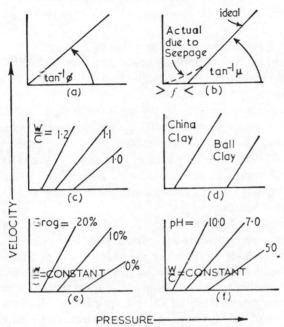

FIG. 64. Plasticity in clays. (a) Liquid; φ is fluidity. (b) Clay: μ is mobility: f is yield value of pressure. (c) Water: clay ratio (W/C) varies. (d) Clay type varies. (e) Proportion of grog varies. (f) pH varies.

reckoned as clay when calculating the ratio). Plasticity is of particular importance in hand moulding where it is desirable to attain good plasticity without excessive water additions which are not easy to dry out. Even in machine moulding good plasticity is desirable in giving more uniform texture throughout the brick.

Manufacture of Firebricks

The manufacture of firebricks follows closely the general description in the last chapter. Drying must be slow enough to avoid cracking, especially of hand-made bricks. Dry-press bricks are easily dried.

Firing follows three stages, as follows:

1. Smoking or Steaming—12–48 hr, from 20–300°C under reducing conditions. Mechanically and colloidally held water is expelled.

2. Decomposition—10–24 hr, up to 900°C in an oxidizing flame. Clays decompose over 500°C, combined water being driven off leaving an amorphous residue sometimes called "meta-kaolinite". Any free α-quartz is converted to β-quartz above 573°C. Carbon and sulphur must be burned out because vitrification in stage 3 will prevent their later escape and leave the brick black-hearted.

3. Full firing—12–18 hr, up to 1200–1400°C. The formation of silicates probably proceeds from 1000°C onwards, and the upper permissible temperature depends on the progress of this vitrification. Silica and alumina are converted to higher temperature modifications and combine to form mullite at temperatures above about 1100°C. The top temperature depends on the proportion of fluxes present. Under-firing leaves the centre friable and weak. Over-firing may induce slumping and certainly causes high susceptibility to thermal shock.

The whole operation takes 3–5 days including cooling and the final structure will be mullite needles in a matrix of glass with any free silica as unchanged quartz or as tridymite or cristobalite.

Properties and Uses of Firebricks

Firebricks are classified by alumina content. Strict definition is by B.S.S. 1902: 1952. Aluminous firebricks contain 38–45 per cent Al_2O_3. (Pure china clay contains 46 per cent Al_2O_3.) Ordinary firebricks contain up to 38 per cent Al_2O_3, the lower limit being defined by the proviso that SiO_2 must be less than 78 per cent—i.e. Al_2O_3 above about 22 per cent. Nevertheless, bricks with less than 32 per cent Al_2O_3 are of low quality and may be referred to loosely as siliceous firebricks. If Al_2O_3 is very low, between 11 and 22 per cent, the brick is called semi-silica and will be discussed later.

The levels at which other oxides may be found are shown in Table 9. Titania can be surprisingly high. Fe_2O_3 may come up to 3 or 4 per cent very readily. Alkalies may rise to 2 per cent but should be less than 0·5 per cent if possible. CaO and MgO usually account for another 1 or 2 per cent between them. It is the cumulative effect of these that is important in forming silicate bond in the firing stage and ultimately in fluxing the brick. They also act with slag in any attack on the brick. Obviously the lower these oxides are the better the brick is likely to be.

TABLE 9

Composition of Silica and Alumino-silicate Bricks

	SiO_2	Al_2O_3	TiO_2	Fe_2O_3	CaO	MgO	K_2O
High-duty silica	96/97	0·5/1·0	0·2	1·0	1·5	0·1	0·2
Silica	94/95	1·0/2·0	0·5	1·5	2·0	0·2	0·3
Semi-silica	88/90	8/9	0·5	1·0	0·2	0·2	0·4
Siliceous firebrick	63/65	26/28	Range	Range	Range	Range	Range
Firebrick	60/62	30/32	0·5	0·7	0·1	0·3	0·1
Aluminous firebrick	50/52	40/42	to	to	to	to	to
Andalusite	36/38	54/56	5·0,	6·0,	1·0,	2·0,	3·0,
Sillimanite	30/32	60/62	1·5	2·0	0·5	0·8	1·5
Mullite	20/22	70/73	typical	typical	typical	typical	typical

Other typical properties of firebricks are listed in Table 10.

It will be seen that the refractoriness rises from about 1550°C to 1750°C as alumina rises from 25 to 45 per cent. Under load, however, softening is always evident at about 1500°C and may be even earlier depending on details of composition. Spalling resistance is usually good, the better again if alumina is high. Reversible thermal expansion (Fig. 60) compares favourably with other bricks.

TABLE 10

Typical Properties of Silica and Alumino-silicate Bricks

	Real density g/cm³ *	Bulk density g/cm³ *	Apparent porosity %	Refractoriness °C	R.U.L.	Spalling resistance	Slag resistance	Cold crushing strength lb/in.²	MN/m²
High-duty silica	2·3	1·8	20	1730	50 lb/in.² at 1600° No deformation	Good above 300°C	Good to acids	5000	34·5
Silica	2·3	1·7	25	1710	25 lb/in.² at 1600° May fail within 1 hr		Fair to FeO	5000	34·5
Semi-silica	2·5	1·7	28	1450	Poor	Good	Good	2000	13.8
Siliceous firebrick	2·7	1·8	30	1550	1200/1300/1400°	Poor	Poor to bases	2000	13.8
Firebrick	2·6	1·9	26	1700	1450/1500/1550°	Fair	Poor to bases	3000	20·7
Aluminous firebrick	2·6	2·0	24	1750	1500/1550/1620°	Good	Fair	5000	34·5
Andalusite	2·8	2·4	22	1780	1520/1570/1640°	Good	Fair	4000	27·6
Sillimanite	3·0	2·3	22	1810	1550/1600/1650°	Good	Good	6000	41·4
Mullite	3·2	2·2	22	1850	1570/1610/1690°	Good	Good to FeO	9000	62·0
Insulating firebrick	2·7	0·95	66	1700	Poor	Poor	Poor	500	3· 5

* SI: To convert to kg/m³ multiply by 10³.

Resistance to acid slags is generally good and resistance to FeO, basic slags and alkalies poor. Slag resistance depends to some extent on texture and is usually better in the higher alumina grades partly because of the alumina and partly because the proportion of alkalies and other fluxes is often rather lower in

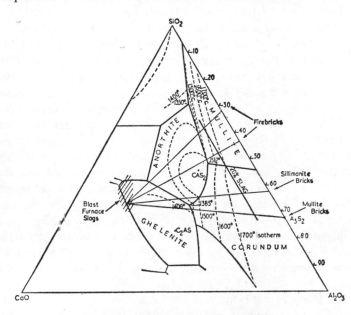

FIG. 65   Part of the liquidus surface of the $CaO-SiO_2-Al_2O_3$ equilibrium diagram (after Greig, Rankin and Wright) drawn to show the relationship between blast-furnace slag compositions and those of various alumino–silicate bricks.

these more expensive bricks. The particular case of attack on alumino-silicates by iron blast-furnace type slags has been thoroughly investigated by Ford and White* using the $CaO-SiO_2-Al_2O_3$ ternary diagram, a liquidus surface of which is reproduced in Fig. 65. This diagram, which is of importance to

* FORD, W. F. and WHITE, J., *Trans. Brit. Ceram. Soc.* **50**, 461 (1951).

both metallurgists and ceramists, has been accurately determined in all areas. The version shown omits some recent improvements toward the Al_2O_3 corner. Compositions corresponding to about 20 per cent of slag and 80 per cent of firebrick all lie in the field in which the primary phase is mullite and the slope of the liquidus surface down to the eutectic trough between the mullite and anorthite fields is so steep that at say 20 per cent slag and 1500°C, even ignoring the contribution of fluxes from the brick itself, about 90 per cent liquid can be expected. Even at the sillimanite composition where corundum separates first, things are no better and the eutectic valley between mullite and anorthite dominates the whole range of possibilities below the mullite composition. Detailed examination of vertical sections indicates that the optimum Al_2O_3 content is about 44 per cent, i.e. opposite the point 1512° on the diagram, at the top end of the mullite-anorthite eutectic valley, where the slope of the liquidus is least and the proportion of liquid formed with, say, 20 per cent of slag at 1500°C is also least (nil on the diagram up to 1512°). Beyond 70 per cent Al_2O_3 it is obvious even from Fig. 11 that 20 per cent of blast-furnace slag could be absorbed without formation of much liquid at 1500°C but mullite bricks are, of course, very much more expensive than firebricks and are not made from indigenous rock.

The effect of FeO on these bricks is important but the FeO–AlO_3–SiO_2 diagram is incomplete. Towers,† however, has discussed the attack of alkalies on firebricks and shows that high-alumina firebrick resists the attack much better than siliceous firebrick by considering the Na_2O–SiO_2–Al_2O_3 diagram. The inherently viscous nature of molten silicates must play an important part in delaying the failure of these bricks under corrosive conditions. Provided the glassy products of the initial attack do not flow off the surface they provide protection against further direct attack and render indirect attack by diffusion a very slow

† TOWERS, H., *Iron and Steel*, **28**, 101 (1955).

process. This is probably more effective in the siliceous firebricks than in the high-alumina ones.

Firebricks are sometimes subject to carbon deposition from CO in furnace atmospheres at temperatures about 450°C (2CO = CO_2 + C). This is catalyzed by "iron-spots" which should be absent. These are derived from particles of iron oxide or sulphide in the clay. Hard-firing converts all iron to silicate, when it is harmless, but must not, of course, slag the brick in the process. Low porosity and particularly low permeability restrict access to this gas. A special test may be necessary to ensure a brick to be proof against this form of failure, particularly when the brick is to be used in the stack of a blast furnace. A sample is held in a stream of carbon monoxide at 450°C for up to 200 hr and examined at intervals for signs of carbon deposition, cracking or disintegrations (see B.S.S. 1902 : 1952).

Hard firing is also a means of increasing abrasion resistance. This is by partial vitrification and consequent reduction of porosity, resulting in a more compact and harder brick. Spalling resistance may suffer. In such a brick high refractoriness need not always be necessary.

Incorporation of grog—prefired clay or crushed broken brick— in the clay mix usually leads to a harder brick with good spalling resistance as advancing cracks must find their way around grog particles. Grog is advantageous in a number of respects in the manufacturing stages too.

Firebricks are the common brick of the furnace designer. They appear in various qualities all over industry where heat is applied as well as in the domestic fireplaces and stoves. They are moderately cheap and the material is indigenous and traditional. Replacement at special points has been steadily going on for forty years but they still dominate the scene.

In ironworks they appear in blast furnace stack, bosh and hearth and in ladles. They fill the Cowper stoves and line the hot-blast main and blast-furnace main. In steelworks they appear in the open hearth but not exposed to the steelmaking temperatures. They fill the checkers, line the ladle, form its stopper, and appear in the casting bay as runners and guide tubes. Again in the soaking pit and reheating furnace firebrick is the basic building material. It appears too in small furnaces in foundries and in non-ferrous industries and in all types of steam-raising plant. The quality used depends on the conditions prevailing and obviously the cheapest that will do the job should be employed. High alumina qualities go into the blast furnace and many of the steel plant applications, e.g. at the *top* of the Cowper stove and checker assemblies where high temperatures or potentially slagging conditions are met. The blast-furnace stack is liable to carbon deposition and here china-clay bricks (no iron), made to very low porosity for high abrasion resistance, have been tried. In ladles a quality which can be relied upon to "bloat" or expand on heating, that is having by design a high after expansion, is used to ensure tight jointing (see Fig. 60). This bloating seems to be due to an evolution of gas (probably SO_2) after the porosity has been sealed off and when the brick has started to soften.

High Alumina Bricks

It is obvious that increases in Al_2O_3 content beyond 46 per cent would be beneficial, but there are no suitable clays in Britain to make such bricks so that minerals have to be imported for the purpose. Sillimanite ($Al_2O_3.SiO_2$), imported from Assam, containing 63 per cent Al_2O_3 was first used in the 1930's, but is today replaced by kyanite from Africa. More recently South African Andalusite (of the same composition as sillimanite and kyanite but a different crystal habit) has been adopted to make a brick of 50 per cent Al_2O_3. These minerals are in the form of rocks and may be used as mined, or blended with clays to obtain intermediate compositions. The terms sillimanite and andalusite are applied to the bricks although the constitution after firing is mullite, glass and cristobalite or tridymite as in the firebrick, except that the proportion of mullite is higher and the amount of fluxes probably lower.

Bricks are made also to 70–73 per cent Al_2O_3, called mullite bricks, and containing very little free silica after firing. Bauxite must be used to raise the alumina content this far. Alumina bricks are made entirely from bauxite with a small amount of clay added as a bond. The bauxite may be pre-fused, crushed and ground, but this makes a very expensive brick. Such a brick is entirely mullite and corundum and remains solid up to 1810°C (Fig. 63). These products are progressively more expensive than firebricks and their use is limited to places where conditions are severe. Table 10 shows how refractoriness improves at least up to the mullite composition, and there is a steady improvement in hot strength through the range, while spalling resistance remains excellent. The cold crushing strength of some of these bricks is extremely high and they are used where liable to heavy mechanical wear or abrasion, and not necessarily always at very high temperatures. Resistance to slagging is generally rather better than that of firebricks. The 50–60 per cent Al_2O_3 bricks stand up to acid slags and glasses, but at 70 per cent Al_2O_3 resistance even to FeO and lime is quite good. None remain unaffected by alkalies but high Al_2O_3 is an advantage here too and these bricks are used in glass tanks.

Each type of these bricks can be prepared to a range of specifications, the control being mainly through manipulation of the porosity. The properties listed in Table 10 are only typical and no attempt is made to indicate the ranges within which the individual values may lie.

The use of high alumina bricks has been extended steadily over recent years, replacing the better qualities of firebricks in many applications. The mullite brick has rapidly become recognized as one of the most durable of refractories. It is now used to line the bosh of blast furnaces where it does actually survive in a recognizable form whereas the firebricks previously used were rapidly slagged back almost to the cooling blocks. Higher up the blast furnace, in the lower stack, sillimanite and andalusite bricks have replaced top quality firebricks which are retained only in the upper stack where they provide the best resistance against carbon deposition provided they are made dense and well fired. Mullite and andalusite bricks are also being incorporated into the upper courses of checker work in regenerators and Cowper stoves. High density bricks are required for this application with porosity about 12 per cent to provide good thermal conductivity and a high capacity for heat. These bricks also appear in soaking pits, reheating furnace walls, roofs and hearth and around doors. Mullite bricks are normal in electric arc steelmaking furnace crowns and in the lower parts of soaking pits where attack by FeO is severe. Non-ferrous metallurgical applications include brass melting reverberatories, lead drossing reverberatories and aluminium smelting furnaces. There are also non-metallurgical applications in glass, ceramics, cement and enamelling as well as in high duty boilers. Mullite is also beginning to replace silica bricks in coke oven walls in a very dense, highly conductive form. The "alumina" (95 per cent Al_2O_3) brick still has a few applications which can justify its cost.

23

Silica Bricks

RATHER pure silica is required for brick-making—not less than 95 per cent SiO_2 and with less than 1 per cent Al_2O_3 and 0·3 per cent alkalies. Ganister, quartzite, sand and flint are all used, and from South Africa micro-crystalline material called silcrete which has TiO_2 as a major impurity. Other sources of suitable material are continuously being sought.

Ganister is clay-bonded sandstone and has been used raw in furnace construction for centuries. In the South Pennines there are ganisters with up to 98 per cent SiO_2 suitable for brick-making. Most ganisters are expensive to mine and quartzite, mainly from Wales, has been the chief source of silica in Britain in recent years, although these reserves are running low. This is a sandstone bonded with colloidal silica and is very hard. Sand may be blended with quartzite, and broken bricks are returned like grog.

The quality of the brick depends on the crystal size of the rock, so that not all quartzites are suitable. Certain types with very fine crystals and silcrete which shows a conchoidal fracture and extremely fine structure, fire to a brick of unusually low porosity, and, other things being equal, enhanced properties. There has been developed in recent years a grade called high-duty silica bricks derived from specially selected materials, high in SiO_2, low in Al_2O_3 and alkalies, and of very fine grain size and low porosity.

Allotropy in Silica

There are said to be 15 modifications of silica. The important

273

forms are set out in Fig. 66 to show their ranges of stability and the volume changes occurring on transformation, according to Sosman.*

The naturally occurring form is α-quartz, which transforms rapidly and reversibly to β-quartz at 573°C. The volume change is 1·35 per cent, large enough to shatter a piece of the material or an unfired brick unless heating past this temperature is very slow. β-quartz is stable to 867°C but may persist to much higher temperatures. Its conversion to tridymite is slow unless catalyzed by a mineralizer such as calcium tungstate, as a considerable degree of atomic reorganization must occur. Mineralizers are usually fluxes and cannot be used in sufficient quantity to effect rapid transformation in bricks. However, cracking is not likely during slow transformation. Above 1470°C cristobalite is the stable form and again the transformations, either from tridymite or from quartz, are slow. Cristobalite remains stable to the melting point at 1723°C.

Sosman's classification shows conversion of quartz and tridymite to glass rather than cristobalite at 1450°C and 1680°C respectively. Liquid, on cooling, forms glass or vitreous silica, which is of course a metastable form, and devitrification to cristobalite could occur in time if a high temperature were maintained and particularly in the presence of a suitable mineralizer.

Cristobalite, when cooled below 1470°C, can revert to tridymite, but this is slow and usually it persists as metastable cristobalite to room temperature but with a complex transformation at 275–200°C to a low temperature variety. Similarly, tridymite persists to low temperatures with modifications at 475°, 210°, 163° and 117°C.

The volume changes involved are rather high for such a low temperature and rather spectacular shattering occurs unless the temperature is lowered very slowly from 300° to 100°C. Similarly,

---

* SOSMAN, R. B., *The Properties of Silica* (American Chemical Society Monograph, 1927).

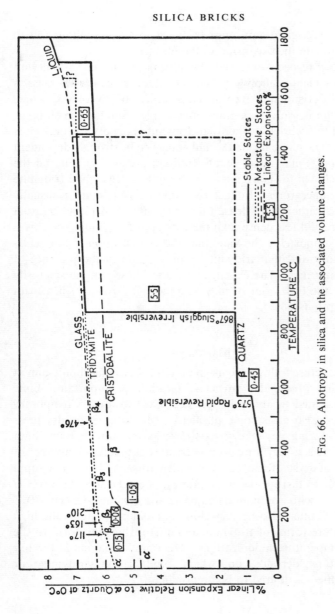

Fig. 66. Allotropy in silica and the associated volume changes.

Note: The line for glass probably ought to dip below the tridymite line in the middle range of temperature but has been drawn above it at all points for clarity.

on reheating, great care must be taken to avoid spalling and to accommodate the expansion in the furnace.

In a few respects observed phenomena are not consistent with Sosman's classification (or Fig. 66). Devitrification of glass is always to cristobalite even at temperatures below 1470°C where cristobalite is, on the diagram, metastable. Similarly cristobalite is found in firebricks rather than tridymite even after firing below 1470°C. The phase formed would appear to be strongly dependent on the particular fluxes or mineralizers present. In Fig. 66 the glass is shown for convenience as being less dense than tridymite at all temperatures. In fact the super-cooled liquid is usually shown as rather more dense than tridymite. It is rather unusual to find a solid less dense than the corresponding liquid and this is further support for the view that the classification is not really satisfactory and that tridymite is only stable in the presence of some impurities. The classification does, however, satisfactorily describe the behaviour of most commercial forms of silica, with the exceptions noted.

Manufacture

Manufacture follows the usual pattern. Crushing should aim at production of angular particles. Grading is important. Lime water is added to about 1·7 per cent CaO to form a temporary bond and also gives some plasticity. Moulding is by dry-press method. Firing may take up to two weeks and should attain about 1500°C. It is a very long process because it involves a number of changes of crystal form, some of which are very slow, some rapid and liable to disrupt the brick (see Fig. 66). Heating past 573°C must be slow to accommodate quartz and cooling between 300°C and 100°C must also be very slow to avoid cracking as the high temperature forms of tridymite and cristobalite transform to the room temperature modifications. The time of soaking at the top temperature depends on which modification—tridymite or cristobalite—is desired to be predominant.

It is usual to fire "hard" so that no quartz is left in a brick. Then after-expansion is low, and all the care has to be taken up to 300°C only, on heating up the furnace. No trouble at 600°C need be anticipated. (Soft firing is adopted in Germany apparently because they prefer a less sudden expansion over a longer period.) The resulting brick is then a mixture of tridymite and cristobalite depending on the firing temperature and time, and the prevalence of suitable mineralizers. Where occasional cooling toward 200°C may occur the brick should be mainly tridymite. During high temperature work, however, gradual conversion to cristobalite is likely, and most bricks will be made with much of that form present when new. The constitution is reflected in the density of a brick, compared with the densities of quartz 2·65, tridymite 2·26, and cristobalite 2·32 g/cm³ (2320 kg/m³).

Bonding of the particles in the bricks is partly by formation of calcium silicate glass and pàrtly through an interlocking action between tridymite needles. The most valuable property of silica bricks is a very high R.U.L. which is due to this tridymite bond. Rigidity can be maintained even under load very close to the melting points. It might seem that 1·7 per cent CaO is rather a powerful flux in such a brick. Figure 67 shows, however, that there is a "monotectic" reaction at 1690°C in the CaO–SiO_2 system below which only a very small proportion of liquid can form— about 5 per cent with 1·7 per cent CaO, down to 1440°C. This has little weakening effect on the strong tridymite skeleton. FeO, MgO and MnO behave in a similar manner (Figs. 62 and 68) with silica but Al_2O_3 and TiO_2 both form eutectics with SiO_2 at about 1545°C with 6 per cent and 10 per cent additions respectively (see Fig. 63). These and alkalies should be kept low in general but TiO_2 itself can be tolerated in some of the high-duty bricks of the silcrete type. This is because the monetectic "platform" extends deep into the CaO–SiO_2–TiO_2 equilibrium diagram despite the eutectic in the SiO_2–TiO_2 system. A rather high CaO content can be an advantage here. Alkalies and alumina in small concentrations cause the monotectic platform to disappear.

Properties and Uses

Refractoriness of silica bricks is often quoted up to 1710°C—formerly adopted as the melting point. Occasionally even higher values are obtained for new high-duty bricks—e.g. 1730°C.

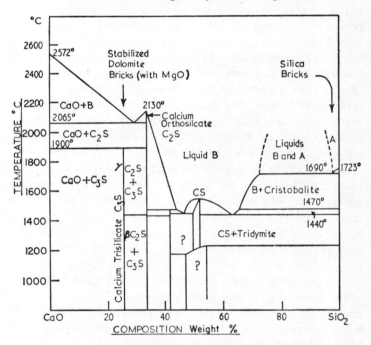

FIG. 67. The CaO–SiO_2 equilibrium diagram (after Rankin and Wright: and Greig) showing the "monotectic" point and wide immiscibility gap which control the liquid content at a low level even in the presence of some CaO. The trisilicate and orthosilicate of calcium used in the stabilization of dolomite also appear in this diagram. *Note the notation:* $C = CaO$, $S = SiO_2$, $C_3S = 3CaO.SiO_2$. In other diagrams $F = FeO$, $A = Al_2O_3$, $M = MgO$.

Under load, initial softening occurs at about 1630°C in ordinary qualities and up to 1670°–1700°C in high-duty bricks. In maintained temperature tests, however, 10 per cent subsidence would occur at 1450°C in an hour in ordinary grades and at 1650°C in 2 hr in the higher quality. The improvement in developing the high-duty bricks is considerable, although the cost is high. The thermal expansion of silica between the allotropic change points

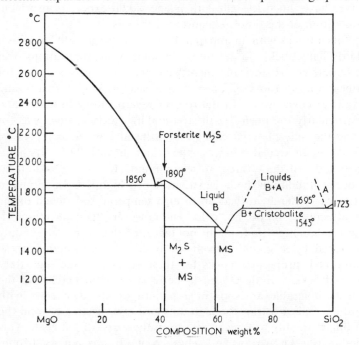

FIG. 68. The MgO–SiO$_2$ equilibrium diagram (after Bowen and Anderson) showing the "monotectic" point on the right-hand side, and also the very refractory nature of forsterite M$_2$S compared with MS.

is very low, and spalling resistance above 300°C is therefore very good, although so poor below that temperature (see Fig. 66). Resistance to slagging is remarkably good, considering the extremely acid nature of the brick. Permeability is usually very low indeed and porosity is held as low as possible i.e. down to about 15 per cent in the high-duty bricks. It is again the form of the $CaO-SiO_2$, $FeO-SiO_2$ and $MnO-SiO_2$ equilibrium diagrams which is mainly responsible, however, for the resistance to slagging, coupled with the very high viscosity of the product of reaction. Not only acid slags, but FeO and to a lesser extent basic slags have much less effect than one would expect from a consideration of the simple chemistry of the case.

Silica bricks with an apparent density of about 1800 kg/m^3 is a fairly light brick—an advantage in some of its applications over alternatives. Historically there have been two principal uses for silica bricks—roofs of open hearth steel-making furnaces and walls of coke ovens. In the roof of reverberatory furnaces, and particularly the open-hearth steel-making furnace, silica was for long the only material capable of withstanding the compressive forces in an arched roof at temperatures up to 1600°C. Its lightness reduced these forces to a minimum. Its low expansion favoured stability in the large structure. Its high R.U.L. permitted it to carry the great load up to high temperatures—much of the load being taken by the "cold" end however. Its high refractoriness minimized drip from the hot face, and that face is not unduly attacked by slag splash from the bath. Notwithstanding these admirable properties it was for these roofs that the high duty silica bricks were developed to be used either in patches at the most vulnerable areas or in a pattern of alternate rows with ordinary silica bricks, the high duty bricks projecting beyond the ordinary quality to screen them from the worst of the heat. This design was known as the "zebra" roof. It was not used long, however, because of the almost simultaneous development of basic roofs.

Silica bricks in open hearth furnaces failed by gradual solution of the hot face by fluxes thrown up from the bath below. Molten silicates then dripped back into the bath below. At the same time the bricks suffered structural changes and developed a banded structure parallel to the hot face. These bands were made up as follows. First there was a hot face layer of dense grey cristobalite ahead of the 1470°C isotherm. This was backed by a darker zone of tridymite. Then there was a narrow zone into which fluxes, including Al_2O_3, had migrated back from the hotter zones. Behind this there was an unchanged zone. The loss of fluxes from the cristobalite zone may cause its refractoriness to be higher than that of the original brick in spite of absorption of FeO from the process, but there is a tendency toward structural spalling through the interfaces between the zones and particularly where the fluxes have segregated. If this zone is exposed by such spalling, lateral attack by slag can be severe.

Other reasons for failure are damage by thermal spalling during heating up, which should not exceed the rate of about 10° per hour between 200° and 300°C, and overheating by, for example, direct flame impingement which causes rapid slagging off and even straight melting of the brick, until the roof is dangerously thin.

The second major application is as the oven walls of coke-ovens. Here a thin wall (for rapid heat transfer) is required to stand at 1400°C almost indefinitely. Silica bricks of high quality were originally developed for this purpose. The monopoly enjoyed by silica in coke ovens is, however, now being challenged by mullite and specially high density modifications of both types of brick are now being developed to improve heat transmission through the walls and also improve the resistance of the brickwork to abrasion by the coke while it is being pushed out of the ovens.

A third important application of silica bricks is in the domes and upper checkerwork of Cowper stoves. This is a relatively new application arising from demands from the iron makers for higher blast temperatures so that the Cowper stove dome temperatures have now risen to about 1500°C. This is an ideal application of silica which will survive the compressive stresses of an arch or dome at that temperature almost indefinitely. In the top checkers a high density type of brick is preferred for high thermal conductivity and high thermal capacity. These are usually made to special shapes designed to present a large specific surface to gases passing through which will optimise heat transfer rates. Hexagonal prisms with hexagonal channels running through them are currently preferred. The effective thickness of the silica in these is only about 3 cm so that the turn-round time of the stoves is very short—about 20 minutes. Silica can be used only at the top of the checkers where the stove temperature remains high all the time. Further down denser mullite is more suitable.

Elsewhere silica can be used only with care in view of its spalling characteristics, for it cannot be cooled below 300°C. It is used in some small acid electric arc furnaces and small foundry cupolas and sometimes in situations where its resistance to slagging is useful as in the bottoms of soaking pits. In general it should be looked upon as a special brick for special purposes.

The acid open-hearth furnace was built largely of silica and the hearth was lined with ganister. Bessemer converters were usually lined with soft-fired silica brick and had firebrick tuyeres. These purposes are now virtually obsolete.

A furnace with silica brickwork is almost impossible to cool below 300°C. This is a major restriction on the use of this refractory.

Semi-silica Bricks are prepared from low-grade ganisters or from artificial sand/clay mixtures. Their properties lie between those of silica and firebricks. They are cheap and have the advantage over firebricks of low after-shrinkage and over silica of better spalling resistance. Slag attack usually looks very severe but is actually restricted to a glazing of the surface. It has been suggested that this brick is always far from equilibrium, remaining always a firebrick matrix with silica particles embedded in it. These bricks are used as cheap backing for silica in places where conditions are not severe or in places where silica would be the choice at a rather higher temperature level. Examples of uses are in coke-ovens, kiln roofs and flues.

Sand Bricks. A recent development in ladle linings is the use of either sand bricks or gunned sand instead of bricks (see page 256). In this application the shell of the ladle is protected by a few runs of conventional brickwork. The sand lining is of a quality rather like semi-silica bricks. It is usually cheaper and quicker to instal than normal brickwork and can be made to last at least as long. The cost of gunning equipment makes that system uneconomical for small units. The thermal conductivity of this lining is low so that heat losses from the ladle so made are minimized. A purer silica sand is used for lining the similarly shaped but smaller induction melting furnace—up to two or three tons capacity. The sand is rammed against a steel former which is heated inductively to initiate firing of the sand. This is completed during the first heat.

24

Magnesite-Chromite Refractories

Magnesite

Lime and magnesia are among the most refractory oxides melting at 2570°C and 2800°C respectively. Unfortunately, lime cannot be dead-burned to become inert to water and its use, as a pure oxide, is restricted to platinum metallurgy.

Magnesia is neither so plentiful nor so cheap as lime. It has traditionally been prepared from magnesite ($MgCO_3$) or breunnerite which contains also some $FeCO_3$. The best deposits are in Austria and Czechoslovakia but Greece, Yugoslavia, India, Australia, South Africa and California all have useful supplies. Russia and Manchuria also produce magnesite. British deposits are poor and here we have turned to the sea as our main source of magnesium, and America also uses the sea for part of her requirements.

Magnesite is prepared if necessary to lower $CaCO_3$ and $FeCO_3$ content by "ore-dressing" techniques—grinding and washing or calcination and magnetic separation—and is then fired in rotary kilns over 900°C to form the oxide and then beyond 1600°C to produce the compact form of magnesia called periclase, which is stable to water. This is known as "dead-burned magnesite".

Sea water contains 0·2 per cent MgO (as $MgCl_2$) so that 110,000 gal or about 500 tons have to be treated to obtain 1 ton of MgO. The chemistry involved is simple. Dolomite is calcined to a mixture of CaO and MgO, called doloma, which is added to the sea water. The oxides hydrate, and $Ca(OH)_2$ reacts with

$MgCl_2$, giving $Mg(OH)_2$ which precipitates while Ca goes into solution as chloride. A number of side processes have to be operated to reduce contamination of the magnesite by impurities in the burned dolomite, and by calcium as the carbonate or sulphate in the sea water itself. Precipitation should be controlled to give a very complete yield of MgO and to produce large crystals which will settle out rapidly in the thickeners. After being separated by filtration, the hydroxide must be dead-burned to periclase as described above.

Brick manufacture is quite straightforward. The clinker is crushed and sized and the grades blended with bonds such as milk of magnesia, hydrated ferric oxide or clay and not over 5 per cent water. Moulding is by dry press at 1000 bar. This high pressure is necessary to give good green strength and fired strength; to minimize shrinkage in firing; and to reduce porosity and improve R.U.L.

The bond which develops is a cement of small crystals of silicates and other compounds. Silica is an inevitable impurity and the desirable compound is forsterite, $2MgO.SiO_2$, which melts at 1900°C. Additions of pure MgO in the finest grades are made to increase the $MgO-SiO_2$ ratio in the bond and hence improve the refractoriness (see Fig. 68). Silicates of iron and calcium and even magnesium ferrite may also form.

In basic bricks refractoriness and strength depend on the properties of this cement, and ultimately on its melting point. The corresponding liquids are not usually viscous. There is no interlocking of acicular crystals, as most of the cement constituents are of cubic habit.

Properties and Uses

Refractoriness of these bricks is about 1800°C, but R.U.L. is not above 1500°C. Spalling resistance is also poor, probably owing to the high coefficient of thermal expansion (see Fig. 60), but some grades now show 30+ reversals. Zirconia is sometimes blended in to this end. "Structural spalling" also occurs due to a swelling caused by absorption of FeO in certain circumstances. Flakes then break off the hot face. Resistance to slags rich in CaO and FeO is extremely good, and even acid slag attack is inhibited by the formation of the high melting forsterite ($2MgO$. SiO_2). Thermal conductivity is very high. Bricks are rather dense, light brown in colour and slightly magnetic.

Magnesite bricks are used in basic electric and open-hearth furnaces, under the hearth and in the walls to some extent. They have been used in roofs, too, and as linings for mixers. Their slag resistance also makes them a possible choice for the top runs of checker work, provided that bricks of high spalling resistance can be supplied.

Magnesite may be built into monolithic walls by metal casing with mild steel which oxidizes and cements adjacent bricks together. Tubing rammed with dead-burned magnesite and assembled into walls before heating up to 1400°C also consolidates into strong structures relatively free from spalling troubles. This type of wall has been used in the kilns in which the bricks are fired, and in electric arc furnaces.

Crushed magnesite can be used for making the working hearth. It is milled to $\frac{1}{4}$ in. for high packing density and rammed in with tar to a template and the tar burned out *in situ*. Similar pea-sized material without the fines can also be used for fettling between heats. In these applications in Britain, however, doloma is more likely to be the economic choice than magnesia.

Dolomite

There are vast deposits of dolomite in Britain. It usually contains an excess of CaO over that in the true double carbonate $MgCa(CO_3)_2$, useful material having about 30 per cent CaO and 20 per cent MgO with CO_2 and varying small amounts of SiO_2 (say 0·5 per cent), Al_2O_3 and Fe_2O_3. It can readily be calcined to mixed oxides ("doloma") but cannot be "dead"-burned. The

CaO is always reactive to water. MgO and CaO form an eutectic at 32 per cent CaO and 2300°C. Before dolomite can be used as a brick the CaO must be stabilized or the brick will crumble in moist air. This is done by combining it with silica. There are two possible silicates—the trisilicate $3CaO.SiO_2$ and the orthosilicate $2CaO.SiO_2$ (see Fig. 67). The former dissociates at 1900°C to form CaO and the o-silicate which melts at 2130°C. The o-silicate has three modifications, γ above 1410°C, β between 1410°C and 678°C, and α below 678°C. The $\beta \rightarrow \alpha$ change involves a 10 per cent volume change which causes "dusting" (material falls to a powder) though this can be inhibited by additions of Cr_2O_3, B_2O_3 or P_2O_5 (which adversely affect slag resistance).

In U.S.A. stabilization is by formation of the orthosilicate (inhibited) but in Britain the trisilicate is favoured, with a small excess of SiO_2, to throw errors on the safer side.

Manufacture

Silica is added as serpentine ($3MgO.2SiO_2.H_2O$) before calcining and the mixture fired in a rotary kiln at about 1600°C to a stabilized clinker, containing periclase, the calcium silicates and aluminates, or ferrates of calcium and magnesium. The clinker is crushed and sized, graded and tempered with 4 per cent water, dry-press moulded at 700 bar, dried and fired at 1400°C. If stabilization is not complete it shows at the drying stage. The finished brick should withstand 24 hr in boiling water without change. These are Stabilized Dolomite bricks. A cheaper product is the Semi-stabilized brick. Shrunk dolomite is mixed with Fe_2O_3 and Al_2O_3 as bonds and then with heavy oils as a temporary bond. The oil has to carry the Fe_2O_3 and Al_2O_3 all over the clinker surface and keep out moisture until the bricks are in the kiln. Firing is at 1400°C and bricks must be put into service at once, unless they are specially prepared for storage with a coating of a tarry material. Under favourable conditions storage up to a year is possible, but the CaO is always active toward moisture.

Properties and Uses

Analyses of the two types of dolomite brick are compared in Table 11 and properties in Table 12. The properties are not very good. Refractoriness is high as in all basic bricks, but under load it falls to 1450°–1550°C for stabilized and 1350°–1450°C for semi-stable bricks. Spalling resistance of the stable brick is often poor, but a spalling-resistant grade can be made; that of the semi-stable brick is moderately good (20 cycles). Slag resistance is poor, attack being mainly through the bond. Doping with fine MgO to make forsterite can improve this property. Semi-stable bricks are again rather better in this respect.

TABLE 11

Composition of Basic Bricks

	MgO	CaO	Cr_2O_3	SiO_2	Al_2O_3	Fe_2O_3	
Magnesite	85/90	1/3	—	2/3	1/3	2/5	Up to
Chromite	15/20	1/2	30/40	4/6	15/20	12/15	0·2%
Chrome–magnesite	38/40	1/2	28/30	4/6	15/20	10/15	MnO and
Magnesite–chrome	65/75	1/2	6/10	3/4	4/6	4/6	0·5%
Stabilized dolomite	40/42	38/40	—	12/15	2/3	2/3	TiO_2 may
Semi-stable dolomite	36/40	48/50	—	3/5	2/3	2/3	occur in
Forsterite	50/60	0·5/1·0	—	30/40	2/3	6/8	all types

The principal applications of dolomite bricks in the open-hearth furnace have diminished as that process has become obsolescent in the 1960s. Most of the L–D or BOS converters which have replaced the open-hearth furnaces are, however, now lined with dolomite in the form of large unfired blocks. These are large special shapes made to fit precisely within the vessel. They are about 60 cms long and are made of crushed well graded doloma temporarily bonded with tar which burns out as the linings are fired *in situ* when put to use. The blocks consolidate to structures similar to those of dolomite bricks but the density is not so high. They seem to give satisfactory service and are not significantly less

Typical Properties of Basic Bricks

	Real density g/cm³	Bulk* density g/cm³	Porosity %	Refractoriness °C†	R.U.L. Rising§ temperature °C	R.U.L. Maintained temperature	Spalling resistance	Cold crushing strength lb/in.²	Cold crushing strength MN/m²‖
Magnesite	3·5	2·8	19/24		1450–1500 1500–1650 1550–1720	Slow failure at 1600°C in all cases	Two kinds one poor‡ one good	10,000	70·0
Chromite	4·0	3·0	22/27	Very high in all cases above 1750°C	1250–1350 1350–1450 1400–1600	The best creep at 1400°C	Poor	7000	48·3
Chrome-magnesite	3·9	3·0	20/25		1350–1500 1450–1600 1600–1730	The best deform at 1600°C	Very good	4000	27·6
Magnesite-chrome	3·6	2·9	20/23						
Stabilized dolomite	3·3	2·5	22/24		1450–1550 1500–1600 1600–1650	—	Poor	8000	55·2
Semi-stable dolomite	3·3	2·5	22/24		1350 1500 1600	—	Good	5000	34·5
Forsterite	3·4	2·6	20/23		1550 1600 1650	Deformation at 1600°C is slow	Fair	4000	27·6

* Bulk density largely dependent on porosity.
† Values seldom quoted.
‡ Controlled by grading and porosity.
§ R.U.L. very variable—depends on bond.
‖ SI: to convert to kg/m³ multiply by 10³.

durable than magnesite used in the same way but they are usually backed up with a safety run of magnesite brickwork next to the shell. It should be understood that these vessels are relined every few days—perhaps after five days' use. This means that the job must be done quickly and cheaply. It also means that small improvements in quality cannot always be put to advantage. Improvements must be big enough to justify changing the relining schedule which is likely to be geared as much to the duration of the working week as to the life of the refractories. When as in this case there will not be any delay in putting the dolomite into service there is no point in using stabilized material. That is reserved for occasions when long-term storage or transportation would make the use of unstabilized dolomite impracticable.

Dolomite bricks have many other uses. The biggest tonnage apart from the L–D converter is probably the electric arc furnace as a partial replacement for magnesite in less severe situations. It also appears occasionally in ladle linings and stopper sleeves. Like magnesite it is also used in crushed form, pea size and tarred for ramming basic hearths and for fettling. Dolomite is seldom better than magnesite but it is usually cheaper and for many purposes satisfactory in service.

Forsterite

This silicate of magnesium has been mentioned as having good refractory properties—see Fig. 68. It is one of the olivenes and occurs in dunite in which it is strongly contaminated with varying amounts of FeO replacing MgO. Olivenes decompose to minerals like serpentine ($3MgO.2SiO_2.2H_2O$) and talc and steatite (both $3MgO.4SiO.2H_2O$), and all of these are used to make forsterite bricks. Very fine magnesia is added to boost the MgO in the bond and the mixture is dry-pressed and fired.

Refractoriness depends upon the iron content, which should be as low as possible, and should be about 1750°C. Under load, the bricks are typical of basic bricks, deforming slowly at any temperature beyond 1550°C. Thermal expansion is low and spalling resistance only moderately good. The refractory is resistant to both acid and basic slags, though perhaps not so much as silica on the one hand and chrome magnesite on the other. Resistance to FeO is also good but it is not better than chrome magnesite. Uses of forsterite are rather specialized and limited. The most important is in copper smelting, but it has been applied to open-hearth back walls with moderate success, in downtakes, and top courses of checker-work. Both the brick and its applications may well still be in a state of development, and there could be a future for a refractory made of the pure compound.

Chromite

The spinels are a group of minerals with the general formula
$$R''O.R_2'''O_3$$

where $R'' = Fe^{++}, Mg^{++}, Mn^{++}$, etc. (but not Ca^{++}),

and $R''' = Fe^{+++}, Al^{+++}, Cr^{++}$, etc.

Hence

$MgO.Al_2O_3$	Spinel
Fe_3O_4	Magnetite
$FeO.Al_2O_3$	Hercynite
$MgO.Cr_2O_3$	Picrochromite
and $FeO.Cr_2O_3$	Chromite

are members of this group. These have all cubic crystal structure and are hard and refractory and all are mutually soluble. Only spinel itself has been used pure as a special refractory. It can be fused with B_2O_3 as mineralizer, and cast in moulds, and Chesters* mentions one application to the linings of induction melting furnaces which can be made impermeable to molten steel from $MgO-Al_2O_3$ mixtures which expand when they combine on heating to form spinel (cf. bloating firebricks for ladles). Chromite, however, gives its name to a type of brick made from chromite ore which is usually a mixture of chromite and picro-

* *See* Bibliography.

chromite. Chromite ores are found in many countries—Greece, Turkey, Rhodesia, Cuba—and vary greatly in composition. Those with highest chromium content go to metal production. Of the remainder those with high MgO and low FeO are used for brick-making.

A typical Cuban ore has 31 per cent Cr_2O_3, 29 per cent Al_2O_3, 18 per cent MgO, 16 per cent FeO, 5 per cent SiO_2 and 0·4 per cent CaO. Obviously this is far removed from the composition of pure chromite.

Chromite bricks are made by standard methods. Bonds may be added—fine lime or hydrated magnesia or (temporarily) tar. Firing is at 1450°C usually on top of magnesite because their low green strength does not permit stacking these bricks in the kiln.

Properties and Uses

The refractoriness of chromite bricks is about 1700°–1850°C, depending on the ore used. Their R.U.L. is poor and they cannot be trusted beyond about 1400°C owing to the possibility of there being $MgO.SiO_2$ in the bond (see Fig. 68). Spalling resistance is very poor, thermal conductivity being about the same as that of magnesite. The colour is distinctively black and the density high at 3 g/cm^2. The only valuable property is an extremely high resistance to attack by either acid or basic slag.

Chromite brick is the "most neutral" refractory and its limited uses all exploit this unique feature. The most important application is as a single neutral course between the basic walls and acid roof of a basic open-hearth furnace. It also appears in the bottom of some soaking pits, protecting firebrick from FeO, and in acid open-hearth furnaces protecting SiO_2 at the ports. Other possible applications, e.g. at slag lines or in ladles, are limited by the poor general properties.

Chrome Magnesite and Magnesite Chrome

The natural development of the chromite brick (which is not

chromite at all, but depends on its MgO content) was the modification of the MgO proportion and condition. Refractoriness is
enhanced by increasing the MgO–SiO$_2$ ratio, especially in the
bond. This is done by adding MgO in the finest sizes only, and
possibly by screening out some of the finest chromite. Hence a
structure is produced made up of relatively large grains of
chromite in a matrix largely of forsterite. Grading becomes very
important and an excess of free MgO in the matrix can only
increase refractoriness. The usual proportioning for chrome-
magnesite bricks is between 70/30 and 60/40 chromite–magnesite
mixture. There are bricks made with much more magnesite—
about 60 per cent—and these are called magnesite–chrome bricks.

Manufacture is by the standard method using dry-press
moulding except that sometimes bricks are chemically bonded
(e.g. with MgSO$_4$) and fired in the furnace instead of in a kiln.
Kiln firing of these bricks has conventionally been to about
1450°–1500°C at which temperature an essentially forsterite bond
develops between MgO from the fines and SiO$_2$ diffusing out of
the chromite grains. A more recent practice is to fire at 1650°–
1700°C where a much more complete solution of the MgO into
the chromite is effected in the solid state so that the chromite
grains are effectively sintered together with some "catalytic"
assistance from MgO, SiO$_2$, FeO and other impurities present.
The choice of impurities obviously becomes very important and
it is for this reason that only certain chromite deposits are suitable
for making these bricks. Most of the suitable ores now come from
the Philippines. Bricks fired to a high temperature are more
finely grained and denser than those fired lightly. Their properties
are generally considered to be superior and they suffer less
severely from structural spalling. Their cost is, of course, somewhat higher.

Typical analyses are given in Table 11. The bricks are dark
brown and very dense (3·0). Porosity varies from 19 to 26 per
cent, the high value giving a rather lighter brick which is an
advantage to the designer of, say, a roof. The bricks have low

cold crushing strength and are liable to edge and corner damage. Thermal conductivity is rather lower than that of magnesite. Spalling resistance is excellent in most of these bricks but if porosity is reduced by grading or by very high moulding pressures, impairment of spalling resistance may result. Refractoriness is high but R.U.L. values are rather variable and, at their worst, not very satisfactory. Even the best are deforming at 1600°C maintained temperature.

Slag resistance is excellent to basic slag and iron oxide. A very large amount of FeO can be absorbed by these bricks before refractoriness is impaired, but the brick must expand, and this can lead to another kind of failure, namely by "growth". Carbon deposition can also occur under certain conditions of temperature (450°C) and gas composition (CO high). This is particularly likely if the Fe_2O_3 content of the brick is high. Carbon also causes "growth" and ultimately spalling.

Growth due to FeO is explained as follows, there being two effects. First, the hot face being at about 1600°C, the matrix may melt as low as 1400°C, especially if high in CaO. Liquid is drawn back into the brick pores by capillarity, re-freezing on the 1400°C isotherm about 4 cm back. Secondly, FeO is readily absorbed in the spinels (the coarse grains) and as they expand as individual particles so also must the brick. This expansion can be accommodated at the hot porous face, but not at the dense zone along the 1400°C isotherm whither the bond has migrated, and cracking occurs at that zone. Flakes 4 cm thick fall off. The process then repeats itself. This is called "structural spalling" and similar processes occur in other basic bricks and to a lesser extent in silica bricks (see page 280). It is aggravated by thermal fluctuations.

The main remedy for structural spalling is to keep the CaO content down (1 per cent), particularly in the magnesite fraction. This prevents liquation of the bond. Very low total lime is not easily attained—especially when sea-water magnesite is being used—but to keep it low in the bond should be easier, as only small proportions o⁻ specially prepared material would be required.

Magnesite–chrome bricks have very similar properties to chrome–magnesite but are rather less liable to structural spalling. These are the most highly developed basic bricks and they are widely used in the basic open-hearth furnace—in the gas uptakes and burner zones and in the back wall and front wall, and, of course, in the roof of the "all-basic" furnace. Roofs may be suspended, sprung or restrained in compression by downward forces applied from the furnace frame against the thermal expansion of the arch. In suspended roofs special shapes are hung on hangers from supports spanning the furnace roof which can be designed to any shape, including flat. The apparent advantage that damaged bricks can be replaced individually is not always a practical proposition as the hot face tends to become monolithic. These bricks are rather heavy for sprung-arch construction and have rather a high thermal expansion coefficient, which would cause the arch to rise and fall with temperature and the hot face, which would quickly become monolithic, to crack. Later designs usually incorporate the restrained arch in which the shape of the roof is under some degree of control. The bricks are held on the arc for which their taper was designed and large-scale cracking is less likely to occur but large temperature fluctuations should be avoided and gas kept on even during fettling.

These bricks may also be used in electric steel furnaces and in the hearths of soaking pits and reheating furnaces in which FeO attack is severe. In non-ferrous extraction they find a place in copper reverberatories, particularly in the roof. Magnesite–chrome bricks are mainly used in roofs and they are now universally adopted in cement kilns.

These bricks can be laid dry if well shaped or cemented together in the usual way, or using steel sheet, placed between them, which oxidizes, attacks neighbouring bricks simultaneously, and binds them together. Bricks are also available in metal cases which serve the same purpose. Building is then merely stacking. This technique can be applied with chemically bonded as well as with fired bricks—for example, suspended roofs on open-hearth furnaces have been made with metal-cased chemically bonded magnesite–chrome brick. The hangers for these may be attached by embedding steel rods in the brick. These will become bonded chemically to the brick material on oxidation partially to FeO at the hot end but will remain available as hooks or eyes at the cool end for suspension.

25

Carbon

CARBON bricks are made from coke or petroleum coke (which is preferred as it has the higher purity), crushed, size-graded and bonded with tar or pitch. The mixture is moulded or extruded and fired at about 1000°C to make the grade called "carbon" or up to 2500°C in a resistance furnace to produce "graphite". In the former, the tar decomposes to a char which bonds the coke particles together. In the latter the crystal structures of the coke and char develop some of the greater perfection of the graphite lattice and grains grow together to give something like the structure of a polycrystalline metal. These materials are stronger than natural graphite and more readily produced in the shapes and sizes required by industry.

Graphite has higher electrical and thermal conductivities than carbon and is rather denser, but it is not so strong at ordinary temperatures. Properties do depend on manufacturing conditions and on the nature and size grading of the raw material.

These bricks have excellent hot strength and, with a high thermal conductivity, good resistance to thermal spalling. They are scarcely wetted by either metals or slags and are not subject to chemical attack except by oxidizing slags and oxidizing gases, particularly air, CO_2 and H_2O, to which graphite is rather more resistant than carbon. The formation of a protective layer of silicon carbide has been suggested to improve this property. This can be done by heating in an atmosphere of silicon tetrachloride. Carbon will also be attacked to some extent by metals which form stable carbides, including iron. Such attack is not likely to be prolonged but would, of course, contaminate the metal.

One of the most important properties of these materials is their excellent machinability which enables complex shapes to be made.

Both carbon and graphite are used as electrodes in arc furnaces particularly but also in electrolysis plant such as that producing aluminium. Graphite is preferred where a high current density is involved for its lower electrical resistivity, but it is considerably more costly. Carbon is also used as an electrical resistor in resistance furnaces, in the form of rods or tubes or as granules. In this case carbon would be preferred for its higher resistivity.

The main application of carbon bricks or blocks is in iron blast-furnaces. These have been lined entirely with carbon but only in the hearth does the practice continue to enjoy popularity, and sometimes only the periphery of the hearth. When they are used across the floor of the furnace the high temperature zone is extended very deep toward the foundations because of the high thermal conductivity of the bricks. Molten metal is able to penetrate a long way into the brickwork and causes trouble with large bears, floating hearths and break-outs. These faults have been countered by use of large blocks carefully machined and set dry with extremely accurate joints. The practice of the mid 1970s is to build the entire hearth of carbon blocks of this kind but to cool these with currents of air blown through ducts built through the hearth. Cold air is delivered centrally and flushes out through radial ducts. There are also the traditional water cooled staves round the periphery of the hearth. The high thermal conductivity of the carbon makes this system effective in imposing a steep temperature gradient in the hearth which freezes the iron which invades the joints before it can advance far enough to do real damage.

Carbon bricks are established as linings for furnaces making phosphorus, calcium carbide, aluminium and magnesium, and for the conducting hearths of some types of arc furnace.

Plumbago ware is made from fireclay and amorphous (natural)

graphite. This is a standard material for crucibles for melting cast iron and other metals. It is not easily wetted and has a high thermal conductivity. It gradually oxidizes so that carbon is lost, and its valuable properties impaired but it is an economical refractory for use under moderately severe conditions.

On smaller scale work graphite has many applications in laboratories and in industry. In powder metallurgy it is used for moulds and plungers for hot pressing. It appears in laboratories as crucibles, as resistors as in the carbon tube or Tamman furnace, and as the heating element or inductor in high-frequency furnaces. It has a wide range of potential uses in rocket engines, in heat transfer systems, and as a moderator in some atomic piles, where very high purity is, however, demanded.

26

Insulating Refractories

ALL refractories are insulating in some degree, but when it is desirable to minimize the loss of heat without unduly enlarging the structure careful selection from a small group of special materials is desirable.

It is necessary to reduce heat flow by radiation, convection and conduction. The baffling of radiation is simple. A single screen will reduce radiation transfer between two points by at least 50 per cent. Two screens will reduce it by over 67 per cent, three by over 75 per cent, and so on. To reduce convective transfer, the free air space around the hot zone should be divided up into small compartments so that circulation in each is limited. To minimize conduction, the hot zone must be surrounded by a substance of low thermal conductivity at its operating temperature. Gases, except hydrogen and helium, have very low conductivities and air is the obvious choice.

The ideal insulator is, then, some kind of honeycomb structure of minute cell dimensions, filled with air and constructed of something with very low conductivity and extremely thin walls. It should also be rigid at the operating (high side) temperature. Up to about 1500°C porous firebrick is available. It is made from fireclay, mixed with hard wood sawdust or chips. The wood burns out on firing so leaving a light porous brick. High refractoriness can be obtained by careful selection of clay. China clay can be used to give the highest refractoriness. These bricks can be cut with a hacksaw and are very useful for making laboratory furnaces. R.U.L. is always rather disappointing.

The conductivity of insulating firebricks (see Table 13) is not so low as that of kieselguhr, or asbestos bricks, but these are not stable to such high temperatures and would be used beyond insulating firebrick—not above about 900°C. At lower temperatures again glass wool, slag wool, vermiculite, or woolly asbestos can be used or laminated, crumpled aluminium foil. These

TABLE 13

Thermal Conductivities of Various Refractories

	c.g.s. units*		
	300°C	700°C	1100°C
Silica	0·003	0·004	0·005
Firebrick	0·002	0·002	0·003
Sillimanite	0·003	0·004	0·004
Magnesite	0·011	0·008	0·007
Chrome	0·004	0·004	0·005
Chrome–magnesite	0·005	0·005	0·004
Insulating firebrick	0·0006	0·0008	0·001
Kieselguhr	0·0003	0·0004	—
Asbestos	0·0001	0·0001	—
Slag wool	0·0001	0·0001	—
Carbon	—	0·005	0·008
Graphite	0·21	0·12	0·08
Carborundum	—	0·03	0·02

* SI: To convert to W/mK multiply by 418·68.

materials are packed in blankets which can be assembled outside the normal brickwork. The modern trend is, however, to put a thick blanket of refractory wool on the inside of the furnace casing when the furnace is built. It is designed that way so that the low temperature insulation can have the maximum effect. Special products are made for this purpose. One type of refractory wool is made from kaolinite. It is very refractory and very white with a high reflectivity for both light and heat. This can be placed behind radiant heaters to throw heat forward toward the burden. This can readily be incorporated into laboratory furnaces. Low temperature insulation need not be so refractory nor so expensive. Slag wool or even glass wool would be satisfactory for most purposes.

Very high temperature insulation or radiation baffling is effected in laboratory furnaces, by refractory powders—alumina, zirconia, or even carbon. The conductivity of the substance in bulk is of little consequence. In all cases it is the porosity or voidage—amount and dimensions—which determines the effectiveness of the insulation, and the greatest difficulty in their use is to maintain the porosity uniform. Very pure material is desirable to reduce the tendency to sinter and so lose porosity. Carbon is, of course, immune from this but could only be used under reducing conditions or in vacuum. Magnesia is not suitable for use under vacuum at very high temperatures because of its rather high vapour pressure.

27

Special Refractories

THIS heading can only cover a miscellany of expensive materials and new processes which have not yet become common.

1. *Electrocast blocks* are made by fusion and casting. Mullite blocks are used in glass tanks. Fused magnesia bricks and mixtures of chromite, bauxite and magnesite fused and cast into spinel bricks are made in U.S.A. and Germany. Porosity is low; R.U.L. is high; spalling resistance is usually poor, but cast magnesite, resistant to thermal shock, has been made. Growth due to FeO absorption is small, because of the very low permeability.

2. *Alumina* can be obtained quite pure from bauxite ore which is a mixture of two hydrates gibbsite $Al_2O_3.3H_2O$ and diaspore $Al_2O_3.H_2O$, usually associated with a large amount of iron oxide from which it is easily separated. Dehydration gives $\alpha-Al_2O_3$ (corundum) while heating above 900°C forms $\gamma-Al_2O_3$. (A β variation can be crystallized out of a sodium carbonate melt.) Fused alumina may be cast to corundum, crushed, ground and graded and formed into laboratory ware, or the prefired γ modification may be used. Ware should be fired at about 1700°C.

Alumina has a melting point of 2015°C which is low compared with that of other special refractories. It can be used up to about 1900°C. The ware may be impermeable to gases if required, but at the expense of some resistance to spalling. Porous ware is less susceptible to thermal shock and should be used where possible. Slag resistance is poor, especially toward FeO and basic slags. Alumina is unaffected by gases, except fluorine, and it stands up well to fused alkalies. The most usual shapes made are tubes and crucibles.

"Alundum" cement is a useful mixture of fused alumina and clay which can be tempered with water and hand-moulded to any shape for laboratory use. It shrinks a little in firing; it is porous and spalls readily, but is cheap and can be used, depending on the grade, up to 1500°–1700°C.

3. *Laboratory alumino-silicates*. Porcelains and mullites can be obtained in tube and crucible form for service up to temperatures ranging from 1200°C to 1650°C. Those for use up to the higher temperatures are usually more sensitive to thermal shock and are usually the most expensive. Tubes which are not impermeable to gases can have very good resistance to thermal shock and give good service where a gas tight system is not required. These tubes are also much cheaper than impermeable tubes. The cost of holding a vacuum at very high temperatures can be very high. The same materials are used as insulators, e.g. on thermocouples and electric heating elements, and as electrical insulators for high voltage. Their constitution is similar to that of sillimanite and mullite bricks.

4. *Laboratory silica* is available as "Vitreosil"—a pure silica, fused and drawn out to tube form and worked like glass into crucibles, dishes, and other shapes. Normally this is translucent but it can be obtained clear like glass. This material has a very low coefficient of thermal expansion—almost zero—and very high resistance to thermal shock. It can be used up to 1100°C or 1200°C but usually fails by devitrification, a process hastened by the presence of even small amounts of alkalies, alkaline earths and particularly certain special mineralizers like sodium vanadate and tungstate. Devitrification is to cristobalite so that no damage results until the silica is cooled below 300°C. Continuous use at up to 1350°C or even higher is therefore possible, but failure would always occur during subsequent cooling to room temperature. In industry vitreosil is used for replaceable thermocouple sheaths for use up to steel bath temperatures.

5. *Beryllia (BeO)* is used very pure. Its melting point is 2550°C but it sublimes above 2000°C and also reacts with even small amounts of water vapour to form a volatile hydroxide above 1650°C. It is not strong at low temperatures but compares well with other materials over 1500°C. Its thermal shock resistance is particularly good, probably because of its high thermal conductivity.

Beryllia is very unreactive chemically particularly toward metals and carbon and is mainly used for crucibles for containing very pure metals, including alkalies, alkaline earths, rare earths, uranium, silicon—and also nickel and iron. It is useful as a radiation shield in carbon resistance furnaces, having itself a low electrical conductivity, as well as being inert toward carbon. Beryllia dust can be dangerous if inhaled even in small amounts and this can make this useful refractory rather difficult to use.

6. *Magnesia* is made into crucibles for melting pure metals but the manufacture seems to be rather difficult and the product rather less satisfactory than some other pure oxides.

7. *Thoria (ThO_2)* melts at about 3300°C, higher than any other oxide. It is not a very rare material but occurs along with rare earths and its purification renders it rather expensive. Thoria is a basic refractory, reacting with silica but not with basic slags. Its thermal shock resistance is poor. It is radioactive. Its main use is as a crucible for pure metals with which its interactions even at very high temperatures are slight.

8. *Zirconia (ZrO_2)* melts at 2677°C. The oxide is an acid and reacts with basic oxides and slags. It is stable, however, to both oxidizing and reducing atmospheres up to 2200°C (when it reacts with H_2 and N_2) and toward most metals. Pure zirconia occurs in two modifications with an inversion temperature at about 1000°C. Another cubic modification forms which is stable at all temperatures in the presence of lime or magnesia with only a slight fall in refractoriness. The material is prepared either fully stabilized with about 5 per cent CaO or only partially stabilized, with slightly different properties. Neither has very good spalling

resistance because if the inversion is fully suppressed the thermal expansion increases. The thermal conductivity is very low, which makes it a good high temperature insulator.

Zirconia is very unreactive toward metals (except titanium) and is particularly useful for containing refractory alloys. It is used as nozzles through which steel flows in continuous casting equipment.

Zirconia is a sufficiently good conductor of electricity at high temperatures that it can be used as the inductor in high frequency furnaces once primed. On the other hand, it is not suitable as a former for a wire-wound resistance furnace.

9. *Zircon (ZrSiO₄)* melts at 2420°C and is also classed as an acid refractory. It has no inversions but is said to dissociate at 1750°C to zirconia and silica glass. It has a low thermal expansion and hence a good spalling resistance, but this property may be lost if the dissociation is allowed to proceed too far.

Zircon has a variety of rather specialized applications—in the glass industry because of its low reactivity toward acid glasses; in foundries as a moulding sand because of its thermal conductivity; in the remelting of aluminium because of its resistance to wetting by that metal; in boilers where its resistance to attack by oil ash is of value. On the other hand, it is very reactive toward other substances—FeO, fluorspar, and some phosphates—and must be used with care.

10. *Silicon carbide (SiC) or carborundum* is made by fusing together sand, coke and salt in an electric resistance furnace. Part of the product is silicon carbide which can be crushed and reconstituted as bricks or other shapes, by refiring to sinter, or after bonding with clay. It can be used up to about 2000°C under favourable conditions but it tends to decompose in air, Si burning to SiO_2. If this occurs slowly a glaze forms which protects the rest of the material but it can start burning very violently if the protection breaks down. Spalling resistance is very good. Silicon carbide is very hard and is used as skids in billet heating furnaces. It can be attacked by FeO, however, and cannot be used in this

way to very h gh temperatures. Silicon carbide is also used in tube form in recuperators and in bars or spiral form as electric resistors—globars or silit rods or crusilite rods.

Many of these special refractories serve only very limited markets but at the same time specialized users of such materials are searching continuously for outstanding materials to help them solve their design problems in nuclear technology, aircraft and rockets. The materials sought have to replace and do better than the best metals already available at higher temperatures than they can reach. A review by Livey and Murray* has recently surveyed the whole field of possibilities—not only oxides but nitrides, carbides, borides and even sulphides. Few of these are likely to find a place in furnace technology and no more will be said of them here.

The fabrication of articles from these special materials follows the normal pattern outlined for ordinary refractories with a few differences n detail and emphasis. Special refractories are usually made from pure compounds rather than raw ores and the incorporation of fluxes for bonding is not usually possible. Any temporary bonds added must be burned out in firing. Special refractories are often formed by slip casting. Very fine grinding to a few microns is necessary and the particles are dispersed in slightly acid water. Such fine material leads to impermeable ware and assists the sintering process by which the particles are bonded together. This diffusion process proceeds very slowly in the solid state except at very high temperatures so firing is carried out at temperatures much higher than is usual with ordinary bricks. There are, of course, other forming processes such as extrusion of tubes (using a temporary bond), or dry pressing prior to sintering as in powder metallurgy.

Perhaps the outstanding failing of these ceramic materials is

* LIVEY, D. T. and MURRAY, P., contributing in *Physico-chemical Measurements at High Temperatures*, by BOCKRIS, J. O'M., WHITE, J. L. and MACKENZIE, J D.

that they are all inherently brittle. Their strength in compression is good. In tension strength may be tolerably good, but in bending and particularly under impact, very poor. This has been ascribed to there being many fewer dislocation sources operative during slip (than in metals) so that higher stresses are needed to move them than in the case of metals. If the required shear stress is greater than the yield stress slip is restricted to a small number of slip bands which are often found to be associated with micro-cracks. When these cracks extend to a critical size they propagate rapidly. Crystals free of such flaws can be bent considerably before breaking. Alternatively, if the number of slip bands is increased (as by bombarding the surface with hard particles before stressing), the ductility can be greatly increased. This explains the brittleness but does not point to any practical way of overcoming it in commercial materials.

PART FOUR

Instrumentation

28

Pyrometry and Control

INSTRUMENTATION becomes more sophisticated as the demands for closer process control and more efficient operation increases. Temperature measurement is used as a basis for controlling fuel input and to an increasing extent other operations are being taken over by controllers too. The reversal of regenerator systems for example, may be coupled with the temperature of gas leaving the chequers. Apart from process control, records of fuel input, air, oxygen and steam usage, are useful commercial data in analysis of costs.

Pyrometry

Temperature can be measured only indirectly in terms of some temperature dependent property of matter. In principle, almost any such property could be used, but in practice a few have been found more suitable than others for scientific purposes. The experienced furnaceman has, of course, his own means, often remarkably successful, of assessing temperature. These are also based on temperature dependent properties of matter sometimes the same ones as are applied to instrumentation. The colour of light emitted from a furnace is used above 500°C where it is only just visible. Above about 1100°C cobalt blue glasses are required and up to 1650°C the judgement of temperature is, consciously or sub-consciously, a determination of the relative proportions of red and blue light emitted. Below 500°C, "temper colours" are used particularly by tool-makers heat-treating steel. Between 200° and 400°C an oxide film forms on a clean surface of the metal whose thickness is characteristic of the temperature and whose apparent colour varies from a pale yellow to a deep blue and finally black. The "interference" colours are due to the fact that the film thicknesses are of the same order as the wavelength of light. Smelters use the viscosity of slag (how far will it drip off a spoon?) and on the blast furnace the temperature of molten iron during tapping is assessed through the height to which the sparks dance (which is a function of silicon content which in turn depends on hearth temperature).

Temperature scales. If the value of a property such as the volume of a gas can be assumed to vary linearly between two recognizable fixed points then it can be used as the basis of a temperature scale in that range. If the change in its value between the fixed points is divided into n steps or degrees the temperature scale is completed. In 1742 Celsius proposed such a scale divided into 100° between the fixed points indicated by the freezing and boiling points of pure water. This scale is the basis of most scientific work today. While it is popularly referred to as the Centigrade scale in Britain, there is an international agreement to call it by the name of Celsius—in either case, °C.

The Absolute scale is derived from the Celsius scale by considering the properties of a body of a perfect gas which obeys the Gas Law $P.V = R.T$ at all temperatures. If its volume were measured at constant pressure, that would be a measure of temperature. It is found by experiment that if the law remained valid the volume would become zero at $-273 \cdot 15$°C and this is taken as the zero of the Absolute scale (0°A). Degrees the same size as on the Celsius scale are used so that 0°C = $273 \cdot 15$°A, and 100°C = $373 \cdot 15$°A.

The Thermodynamic scale was devised by Lord Kelvin in 1848 in consideration of the working of a perfect heat engine on a Carnot cycle. The zero on this scale (0°K) is that temperature at which the heat engine is capable of completely converting heat into work. When the size of the degree is put the same as the Celsius degree the Absolute and Thermodynamic scales turn out to be identical. The advantage of Kelvin's approach was that the notion of temperature could be divorced from all dependence on the properties of matter, and it remains the fundamental scale defined by the single fixed point, the triple point of water at 273·16°K, now expressed in S.I. as 273·16K.

The Fahrenheit scale has 180° between the ice and steam points and the zero was the lowest freezing point that its inventor could attain with mixtures of water and salts. This put 32°F equivalent to 0°C. The absolute zero on the Fahrenheit scale is at −459·7°F. This scale continues to be used in industry in Britain and in the United States of America and for some purposes in Germany also. Its replacement by the Celsius scale seems inevitable but as it goes so must other units like the British Thermal Unit which are based upon it so that the death may be a lingering one.

The accurate extrapolation of these scales to high temperatures has not been easy. For practical purposes we rely on a number of fixed points on the Celsius scale determined by elegant techniques in gas thermometry and agreed internationally at the General Conference on Weights and Measures in 1948, but always subject to further revision.

The primary standard fixed points under a pressure of one standard atmosphere are:

The boiling point of oxygen	−182·70°C
The freezing point of water	0°C
The boiling point of water	100°C
The boiling point of sulphur	444·600°C
The freezing point of silver	960·8°C
The freezing point of gold	1063·0°C

For interpolation between these points the platinum resistance thermometer is prescribed up to 630·5°C, a standard platinum–platinum–rhodium thermocouple from there up to 1063°C and a standard optical pyrometer beyond that. Interpolation techniques are discussed in standard physics textbooks.

A secondary list of fixed points obtained by such interpolation is also recommended by the same conference and includes the freezing points of the following metals:

Mercury	$-38\cdot87°C$	Nickel	1453°C
Tin	231·19°C	Palladium	1552°C
Lead	327·3°C	Platinum	1769°C
Zinc	419·5°C	Rhodium	1960°C
Aluminium	660·1°C	Iridium	2443°C
Copper	1083·0°C		
(under reducing			
atmosphere)			

Using these values thermometers can be standardized for use in any temperature range.

The important thermometers in furnace technology are the thermocouples and optical and total radiation pyrometers. Before these are considered in detail something should be said about other techniques which are less frequently employed.

Gas Thermometers are important mainly as a means of carrying out fundamental measurements. Equipment is simple. Volume is usually held constant and pressure variations measured but it can be done the other way round. The great difficulties are in making allowances for the expansion of the containing bulb and for its deformation under pressure, expansion in the connecting pipe and gas adsorption. At high temperatures the bulb is particularly liable to creep. Ceramic bulbs are usual but at high temperatures a platinum–rhodium alloy has been used, the pressure inside and outside being kept equal.

Liquid Thermometers are, of course, familiar to all. Mercury in glass or in steel is available to about 600°C under pressure to prevent it from boiling. The range of liquid thermometers can be extended to about 800°C by the use of gallium but lack of

stability in the bulb dimensions makes this type of instrument unsatisfactory for use at high temperatures and even at ordinary
temperatures the mercury in a glass instrument has to be subjected
to stability tests when used for very accurate work.

Resistance Thermometry is a particularly accurate technique
when carried out with all the refinements. Fine platinum wire is
wound bifilarly on a glass, quartz or mica cruciform former and
sheathed in glass or quartz. The leads are brought out through
mica discs which insulate them and act as convection and radiation
baffles. Compensating leads are also brought out on which to
measure the resistance of that part of the circuit which is not in
the head of the thermometer. The resistance is measured using a
modified Wheatstone Bridge circuit full precautions being taken
against stray thermo-electric e.m.f. or heating effects from currents
in the measuring circuit. The tool is rather delicate to be used on
plant in that form but more robust versions using platinum up to
1600°C and tungsten to even higher temperatures have been
developed in the interests of extremely fine temperature control
at high temperatures. These would only be used for very special
work.

Thermisters are made from certain oxides and spinels in the
form of small discs or beads which can be used as very sensitive
thermometers because their electrical resistances decrease rapidly
as their temperatures rise. Unfortunately, they cannot be used
beyond about 300°C because of ageing effects and instability. The
main advantage of the device, apart from its sensitivity, is its
compactness.

Pyrometric Cones or Seger cones are made from mixtures of
oxides and are designed to melt within narrow temperature ranges.
They are used in the ceramics industries either for indicating furnace
temperatures or for testing the refractoriness of bricks. There is a
long series of these cones and an appropriate sequence is set out
together with a specimen of the brick to be tested. The cone which
bends over simultaneously with the brick sample provides the
Cone Number of its refractoriness or this can be translated into

degrees Celsius and called the Pyrometric Cone Equivalent (P.C.E.) temperature. Cones can be used once only. They are not cheap and their use on furnaces is now rare except over 1650°C where the use of thermocouples becomes difficult. Their upper limit is 2000°C. In the test mentioned above they retain the advantage that neither the cone nor the brick have sharp melting points and the collapse is partly regulated by heating rate but if they are compared under identical conditions the effect of heating rate is minimized.

Other Chemical Tests employ chalks and paints of substances which undergo permanent colour change at specific temperatures through decomposition or oxidation. A number of different marks might be made on, say, a casting before it was annealed and perhaps repeated at different parts of it. Subsequent examination would show the temperatures that had been attained in each part. The results would be of particular interest when compared with the temperature of the furnace as indicated by, say, a single fixed pyrometer.

Thermocouples. When two wires of different metals are joined at each end to form a closed loop and the junctions are brought to different temperatures a small electric current can be demonstrated to flow in the loop. More usually if the circuit is opened an e.m.f. can be measured in it which is characteristic of the metals and the temperature difference between the junctions. This was first observed by Seebeck in 1822 but was later resolved into two separate effects. The major one, the Peltier effect (1834) appeared first simply as the converse of the Seebeck effect, namely that if an electric current is passed round such a circuit when the junctions are at the same temperature, heat is evolved at one of them and absorbed at the other, at a rate dependent on the current strength I so that

$$dH_P = \pi . I . dt \qquad (28.1)$$

t being time and π the Peltier coefficient. The minor effect was noticed by Thomson only in 1854. It applies to a single wire in a temperature gradient, in which it is found that an electric current flows (in a closed circuit of which it is a part), or an e.m.f. is set up which may be either in the same direction as the heat flow or opposed to it. The converse is also true in which case

$$dH_T = \sigma \frac{d\theta}{dl} . I . dt \qquad (28.2)$$

where $d\theta/dl$ is the temperature gradient and σ the Thomson coefficient. While π has a value in the range $0\cdot1$–$3\cdot0$ m.cal/C, values of σ are about a thousand times smaller. In practice the current and heat evolutions are never measured but only the net e.m.f. on open circuit.

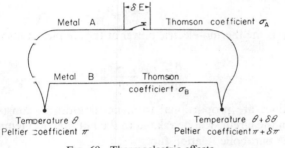

FIG. 69. Thermoelectric effects.

The relationship between the e.m.f., π and σ can be demonstrated. Consider the circuit in Fig. 69 and let the key be closed so that δq e.m.u. of electricity are passed round it the corresponding fall in potential being δE. Applying the first law of thermodynamics that heat absorbed must be equal to work done,

$$[(\pi + \delta\pi)\delta q - \sigma_A\delta\theta\delta q] - [\pi\delta q - \sigma_B\delta\theta\delta q] = \delta E\delta q$$

i.e.

$$\delta E = \delta\pi - (\sigma_A - \sigma_B)\delta\theta$$

or, as $\delta\theta \to 0$

$$\frac{dE}{d\theta} = \frac{d\pi}{d\theta} - (\sigma_A - \sigma_B) \qquad (28.3)$$

Applying the second law of thermodynamics, the change in entropy on completion of the cycle must be zero, i.e. the change in entropy due to the Peltier effect must equal that due to the Thomson effect so that

$$\frac{(\sigma_A - \sigma_B)\,d\theta}{\theta} = \frac{\pi}{\theta} - \frac{\pi + \delta\pi}{\theta + \delta\theta} = \frac{d(\pi/\theta)}{d\theta} \qquad (28.4)$$

Combining (28.3) and (28.4) and carrying out the differentiation of the right hand side of (28.4)

$$\frac{dE}{d\theta} = \frac{\pi}{\theta} \qquad (28.5)$$

or

$$\pi = \theta.P$$

where P is the thermo-electric power at the couple. It also follows that

$$\frac{d^2E}{d\theta^2} = \frac{\sigma_A - \sigma_B}{\theta} \qquad (28.6)$$

If σ values are proportional to θ, i.e. $d^2E/d\theta^2$ is constant, the e.m.f. in the circuit will be related to the hot junction temperature T by the parabolic relationship

$$E = a + bT + cT^2 \qquad (28.7)$$

the cold junction being held at some standard temperature. In fact, such a relationship is widely found to be satisfactory and the linear approximation is often quite close over fairly wide ranges of temperature, though in some couples more complex relationships do exist.

Comparing various metals against platinum as a standard the constants a and b assume values of the order of 100 to 1000 for a the unity for b. No simple system relates the thermo-electric powers of the elements vs. any standard metal. The highest values vs. platinum are given by amphoteric elements like silicon and bismuth but some alkali and transition metals also give high values.

The thermo-electric power is affected by close approach of either element to its melting point and by deformation of the wires.

A wide variety of metal pairs have been used or suggested for use as thermocouples but the selection which might be regarded as standard in industry is quite small. The qualities required of a couple are that the thermo-electric power should be reasonably high and that the e.m.f. produced should increase smoothly with rising temperature. In some cases a maximum occurs, which is inconvenient. Allotropy or phase changes in either wire would cause discontinuities in the relationship (28.7). The wires and their alloy where they are welded together should be stable up to the working temperature in the working environment. They should also be reproducible in quality so that there are negligible thermo-electric differences between one batch and another.

The metal pairs in common use are divisible into the groups, base metal couples and precious metal couples, but there is a growing class of unusual couples for special purposes which falls outside both categories.

Of the base metal couples copper–constantan (60% Cu–40% Ni) and iron–constantan are seldom seen nowadays on furnaces. They had particularly high thermo-electric powers but neither was resistant to oxidation and the former was restricted to about 530°C because of a structural change. They are both very suitable for low temperature work however and copper–constantan is especially suitable for use below 0°C.

The chromel (90% Ni–10% Cr)/alumel (95% Ni–5% Al(Si, Mn)) couple is a most universally used for purposes up to about 1200°C. It satisfies all the required conditions except that it does oxidize slowly beyond about 1000°C and more rapidly at higher temperatures. Its thermo-electric power is good and nearly constant the e.m.f. at 1000°C being about 41 mV., and it is available as two highly reproducible standard alloys. A high proportion of industrial instruments are designed to use it on the assumption that the couples will remain the same for a long time to come.

When temperatures higher than about 1200°C are to be measured precious metal couples are required and here again a standard couple has been developed, namely platinum–platinum–13% rhodium. This replaced an earlier couple which had only 10 per cent rhodium in the alloy wire which was found to owe part of its e.m.f. to a small amount of iron in the rhodium. When the iron was removed the rhodium content was stepped up to 13 per cent to restore the thermo-electric power to its old value. Although this occurred forty years ago, the 10 per cent alloy couple is still occasionally met and is used in the U.S.A. The standard couple can be used up to about 1650°C but is liable to thinning by volatilization, and contamination of the pure platinum wire by rhodium diffused along the couple or volatilized across from one wire to the other. This can be minimized by using a Pt–1Rh/Pt–13% Rh couple for very high temperature work in place of the standard one at some small cost in thermo-electric power.

Modifications of this couple which enhance its thermo-electric power have usually caused a deterioration in stability and a reduction in maximum working temperature. A Pt/Pt–10% Ir couple gives 15 mV. at 1000°C against 10 mV. on the standard Pt/Pt–13% Rh couple but the volatile iridium oxide forms and the calibration is not maintained at high temperatures. Rhenium offers similar advantages and disadvantages. The most successful development seems to be the couple Pt–20% Rh/Pt–40% Rh. Its e.m.f. at 1000°C is only 1 mV. and that at 1500°C about 3 mV. but it can be used to 1900°C in an oxidizing atmosphere.

Tungsten and molybdenum have for long been coupled for use in reducing atmospheres although the e.m.f. developed was very low and actually showed a maximum and then changed sign in the working range. They are now joined by rhenium and tantalum for use in reducing conditions or under vacuum and some of these couples are quoted as being available over 2500°C. The couple W/75%ᵀW–25% Mo in particular has been suggested for use to 2730°C. The coupling of graphite and silicon carbide is useful only to 1550°C but graphite–boron carbide is said to go a thousand degrees further, and graphite–clay "alloys" have been proposed to go a thousand degrees further still. Obviously such instruments are not suitable for everyday use, but rather for research work where great care can be exercised each time a reading is taken and frequent calibrations are possible.

The arrangement of the thermocouple in operation is that the "hot" junction is welded together and the wires are insulated from each other with refractory beads or sleeves. This part is enclosed in a metal or ceramic sheath, suitable for insertion into the furnace, affording sufficient chemical protection for the wires but not so much as to insulate the hot junction thermally from the furnace. The free ends of the wires are brought to the "cold junction box"— a small chamber thermostatted at a low temperature which may conveniently be a Dewar flask with melting ice in it or a small furnace controlled at a temperature a little above the surroundings. The wires are not joined here however. This is the point at which the circuit is opened and the leads (of copper) to the millivolt-meter or potentiometer are taken from it. It is customary under factory conditions to use only short lengths of thermocouple wire and to interpose between the couple wire proper and the cold junction box (which might be at a considerable distance), "compensating leads". These are a pair of wires, made up as twin flex, having exactly the same thermo-electric characteristics at low temperatures below say 100°C as the couple wires but costing a lot less per foot. They are used simply to extend the couple wires with characteristics unchanged from the furnace site where it may be quite hot, to the vicinity of the meter. In many cases the controlled cold junction is dispensed with and the instrument set to a false zero to allow for the fact that the cold junction is not at the standard temperature and this must be adjusted regularly. If the standard cold junction temperature is 0°C (as is usual) the false zero should be the number of millivolts corresponding to the ambient temperature.

The e.m.f. may be read or recorded on a millivoltmeter or a potentiometer or more usually the instrument will be scaled for direct reading of temperature. Millivoltmeters do draw some current and are designed and calibrated for use with some specified external resistance for which adjustments must be made in designing the circuit. Potentiometers take no current and are generally to be preferred but are more costly. The instruments should be housed in a clean, air-conditioned room and should be given daily inspection, particularly if they are effecting control.

Industrial thermocouples are robust and well mounted. The common couples in use are the chromel–alumel and the platinum–platinum–rhodium. The former are made from stout wire or even from a tube of one of the elements and a rod of the other passing down through the tube and insulated from it except where they are welded together at the (hot) end. The precious metal couples are too expensive to allow heavy gauge wire to be used but with probes designed to give fine wires protection and support, and with arrangements for the salvaging of contaminated tips they need not be uneconomical in use. The temperatures of steel baths are regularly taken with this type of couple at the cost of about three inches of each wire and a small silica sheath per dip.

Calibration of thermocouples vs. primary or secondary standard fixed points is effected by obtaining a series of cooling curves of temperature against time as each of a number of substances cools through its transition point. At the gold, platinum or palladium points the hot junction would be bridged with a fine wire of the appropriate metal. At the melting point this wire would melt and break the circuit. Standardization against a carefully calibrated couple reserved for this purpose is an easier way of establishing accuracy. With modern standard couple wire the maker's calibration points are usually good enough for routine work, or would be checked at a single temperature only.

A thermocouple indicates the temperature of its own tip. The value of this temperature depends on the rates of gains and losses of heat suffered by the tip within its sheath by radiation to and from the surroundings, by convection to or from the atmosphere or flame and by conduction along the probe itself and anything else it touches. Unless the temperature in the furnace chamber is quite uniform the indicated temperature must be some sort of compromise—and it may be a most unsatisfactory compromise if, for example, the sheath is licked by flame (convection), or if the couple finds itself close to a cold ingot or a water cooled door (radiation). To minimize these difficulties the probe must be carefully sited and radiation screens may be fitted round it. There is not necessarily a complete solution however and the operator must be clear in his mind that the indicated temperature is probably not the actual temperature of anything except the thermocouple hot junction, but is a compromise value related to furnace conditions in a potentially useful manner. When the actual temperatures of furnace gases or flames are required the special techniques of suction pyrometry and flame temperature measurement by radiation pyrometry must be employed (see below). The accurate measurements of the temperatures of solids and liquids is much more satisfactory as the couple can be inserted in a cavity in the solid or welded to it, or immersed (in its sheath) in the liquid. Contact pyrometers are thermocouples, the hot junctions of which are made from flat strip and mounted like a spring, convex toward the surface whose temperature is required, so that good thermal contact with the surface can be obtained.

Suction Pyrometers are designed to measure the temperature of a gas rather than that of the furnace or flue through which it is passing. The thermocouple is screened from the surroundings to reduce interchange of radiant energy to a minimum, and a stream of the gas is drawn across it or the sheath in which it is enclosed, with very high velocity in order to effect very efficient convective transfer. The screens are also brought near to the gas temperature by the same means. At high temperatures, over 1000°C, the screens may be of extruded ceramic material or may be a bundle of fine refractory tubes, and the head of the assembly may with advantage be water cooled. Suction may be applied by a fan or by steam or compressed air injection. An optimum gas velocity over the tip of the couple of about 200 m/sec has been recommended because acceleration of gas to higher velocities causes a noticeable fall in temperature as some heat is translated into kinetic energy. Suction pyrometers, especially those for use at high temperatures are large, heavy pieces of equipment, mainly because of the water cooling jacket which is necessary to bring the gas temperature low enough for the fan or injector to work efficiently. They are best suited to be permanent fixtures but must be readily demountable for frequent repairs to the screens and couple. The efficiency of the pyrometer should be such that the error obtained using a simple sheathed couple should be reduced by at least 95 per cent, and there are a number of procedures described in the literature for checking that this is being achieved.*

Radiation Pyrometry. Any measure of the intensity with which a black body (or one whose emissive power is known) is radiating heat is also a measure of its temperature. There are two types of instrument—the total radiation pyrometer which uses mainly infra-red radiation and can be used in principle at any temperature and the optical pyrometer which uses filtered red light and can only be used when the temperature is above 500°C at which solid objects first become visible by their own radiation.

The Total Radiation Pyrometer is a device whereby the energy radiated from a part of the surface of the object whose temperature is to be measured is focussed optically, using lenses or preferably by reflection from an untarnishable metal spherical mirror, on to a thermocouple, thermopile or thermistor bolometer which will measure the intensity of the radiation in the image. The optical arrangement should be such that, provided the detector is completely covered by the image of the hot body, the intensity measured is independent of the object distance. (Doubling the distance reduces the amount of radiation received by the instrument by a factor of four but the image area is also reduced to one quarter so that the image intensity is unaltered.) The detector element should be very small and should come to a steady temperature very quickly. If the temperature of the detector is

* LAND, T., and BARBER, R., *J. Iron and Steel Inst.* **184**, p. 269 (1956).

measured by an instrument which gives a linear response D then

$$D = c(T^4 - T_p^4) \qquad (28.8)$$

where $T°K$ is the temperature of the object being measured, $T_p°K$ is the temperature of the pyrometer and c is a constant characteristic of the instrument. If $T \gg T_p$,

$$D \simeq cT^4 \qquad (28.9)$$

and if the object has an emissivity ε,

$$D \simeq c.\varepsilon.T^4 \qquad (28.10)$$

Calibration is essentially the determination of the constant c and this would be carried out by sighting on an artificial black body— a small hole leading from a relatively large cavity in a carbon block would be ideal. The image of the hole must of course cover the detector. The temperature of the carbon block would meantime be measured by a standard thermocouple. The "blackness" of the hole is due to the fact that any radiation falling on it is almost certain to be absorbed within the hole before it can re-emerge. The absorptivity of the hole is therefore unity and the emissivity is unity also. A close approximation to a black body condition is also afforded by a deep wedge shaped cavity.

Once the constant c has been evaluated it should be applicable to the measurement of any temperature above or below the calibration point. At very high temperatures a rotating sector can be used to cut down the radiation received to that which the instrument can handle. If the fraction of the total radiation admitted is f the calibration has to be altered to

$$D = c.\varepsilon.f.T^4 \qquad (28.11)$$

In use the total radiation pyrometer must either be directed on to a cavity such that the emissivity can be approximated to unity or an estimate of the emissivity of the surface sighted must be available with which to apply the necessary correction. A common device is to direct the pyrometer into a silicon carbide tube whose hemispherical end is in the furnace at the point whose temperature is required. The conditions inside are nearly black and possible interference by smoke or flame is eliminated. The whole thing can be sealed off as a unit to keep out dust and ease maintenance. This type of unit suffers the same kind of uncertainties as the thermo-couple probe mentioned above. A further development applied to molten steel is to blow a bubble on the end of the sighting tube with inert gas and sight on that as an almost black body (in spite of its high reflectivity). Surface temperatures can be measured using a hemi-spherical reflector facing down on the surface and focussing the radiation on a thermopile inside—even below red heat.

In the *Optical Pyrometer* an image of a part of the surface of the object whose temperature is to be measured is brought into focus in the same plane as the filament of a lamp the current through which can be varied to bring it to any desired temperature. The image and filament are viewed through an eyepiece and a red filter which restricts the waveband used to about 6500–8000Å the sensitivity of the eye providing the upper cut-off. In use the filament current is altered until the brightnesses of the filament and the image are the same so that the filament "disappears" on the background of the image. These instruments are frequently referred to as "disappearing filament" pyrometers.

Referring to Wein's Law (see p. 167), we know that for a black body at a standard temperature T_1,

$$E_1 = C_1\lambda^{-5} \exp(-C_2/\lambda T_1) \tag{28.12}$$

C_1 and C_2 being constants. For another black body at another temperature T_2,

$$E_2 = C_1\lambda^{-5} \exp(-C_2/\lambda T_2)$$

and therefore

$$\ln\frac{E_2}{E_1} = \frac{C_2}{\lambda}\left(\frac{1}{T_1} - \frac{1}{T_2}\right) \tag{28.13}$$

The most fundamental mode of operation would be to match the brilliance of the filament in the selected waveband with a body at the standard temperature T_1 and then find the setting of an aperture diaphragm which would give a match against a second body at an unknown temperature T_2. A rotating sector diaphragm would be preferred. If it was full open for T_1 and open only to an angle θ for T_2 then

$$\frac{E_2}{E_1} = \frac{2\pi}{\theta} \tag{28.14}$$

C_2 is known to have the value 1·432 cm. degree, so that, inserting a value for λ, in (28.13), T_2 can be calculated. In practice the aperture is maintained constant and the current through the filament of the lamp is varied to obtain a match, the value of the required current being related to the temperature. This renders the calculation less direct. If the resistance R were constant we would have $T^4 \propto I^2$ where I is the current in the filament but as R is usually an unknown function of temperature the instrument must be standardized at a series of temperatures against a thermocouple using a synthetic black body as described for the total radiation pyrometer. In some instruments the filament current is measured indirectly by means of a potentiometer which actually measures the volts drop across a constant resistance in series with the filament, against the potential of a standard cell. There are many designs of optical pyrometer varying in the details of their optics and the method of measuring the current. Their accuracy depends to some extent on the operator but after some practice a single operator should be able to get readings to agree within a range of 3° or 4° at 1500°C and several operators should agree within ±5°. The main difficulties in use are finding a suitably "black" target and avoiding interference by flame or smoke.

Flame Temperatures can be measured in a variety of ways[15] most of which have the disadvantage that they cannot readily be employed in working furnaces without being stripped of some of their refinements. The use of a hot wire in a flame has been suggested. Its temperature can be calibrated against the value of an electric current passing through it when it is held in a vacuum and heat losses are by conduction to its supports and by radiation to its cold surroundings. When in a flame at the same temperature the heat losses are the same as when in a vacuum because the net exchange with the flame is zero and the same calibration ought to be valid but if the flame is in a furnace the surroundings are no longer cold and the calibration is lost. At the same time interaction between the wire and the furnace gases, both due to catalytic effects at the surface and due to interruption of the gas flow pattern introduces uncertainties.

A method employing radiation pyrometry involves three measurements. H_1 is the radiation received from the flame alone, that is with a water cooled target placed behind it on the sighting line. H_2 is the radiation received when the cold target is replaced by a black body at T_b, and H_3 is the radiation received from the black body alone. If ε is the emissivity of the flame and T_f is its temperature, we have

$$H_1 = \sigma \varepsilon T_f^4 \tag{28.15}$$

$$H_2 = H_1 + (1 - \varepsilon)H_3, \qquad (\varepsilon = \alpha) \tag{28.16}$$

and

$$H_3 = \sigma T_b^4 \tag{28.17}$$

whence

$$\varepsilon = \frac{H_3 + H_1 - H_2}{H_3} \tag{28.18}$$

and

$$T_f = \sqrt[4]{\left(\frac{H_1 H_3}{\sigma(H_3 + H_1 - H_2)}\right)} \tag{28.19}$$

The disadvantage of this method is that it assumes that the emissivity of the flame is the same at all wavelengths and that its effective emissivity is always equal to its effective absorptivity toward the heat from the background whatever its temperature. One way of overcoming this is to provide a means of altering the temperature of the black body in the background and to take a series of readings of H_2 and H_3. From relationships (28.15), (28.16) and (28.17),

$$H_2 = H_3 - \sigma\varepsilon(T_b^4 - T_f^4) \tag{28.20}$$

so that H_2 and H_3 become equal when $T_b = T_f$. The point of equality can be discovered graphically and particularly if simultaneous readings of the background temperature can be had from a standard thermocouple, T_f is readily obtainable. The main difficulty is the designing of a suitable controllable background within a working furnace. Readings H_3 would be made through a water cooled tube to screen off the flame.

Of other methods suggested the calorimetric method is probably the most practicable. A sample of gas is drawn at a known rate through a heat exchanger type of calorimeter. The sensible heat in the gas is measured directly and the temperature can be calculated if a specific heat value is available. The greatest practical difficulty would be the transfer of the gases to the calorimeter sufficiently rapidly that the heat loss during transfer would be negligible.

It will be obvious that most of these methods are more likely to provide accurate information about gases or flames under laboratory conditions than on an industrial furnace. This is equally true of other methods not described including spectroscopic techniques. Success on the furnace would be best achieved by taking the refinements of the laboratory on to the plant, as far as possible.

Gas Analysis

It is customary to use flue gas analysis as a means of assessing the efficiency of combustion, and as a check on the influx of air through the brickwork. Analyses on spot samples taken with a gas burette in the standard manner are not really satisfactory for this purpose and it is now possible automatically to obtain a record of analyses with respect to carbon dioxide, the CO/CO_2 ratio, hydrogen or oxygen. Various physical and chemical principles are employed. Carbon dioxide can be recorded by taking a standard volume of gas, passing it through an absorbent reagent and recording the reduced volume. A second absorption and a second measure of volume would provide the carbon monoxide record. Another method employs differential absorption of infrared radiation a standard gas sample being compared with the current sample. Hydrogen is usually measured by a hot wire method. Since the thermal conductivity of hydrogen is much greater than that of any other industrial gas, small changes in hydrogen content affect the temperature of an electrically heated wire and hence its resistance. The hydrogen content of the gas is then reflected in an electrical measurement. Apparatus is also available for recording the oxygen content of flue gas using its magnetic properties. Oxygen along with only nitrous oxide and nitrogen peroxide among the common gases is paramagnetic—that is they are attracted by a magnet. It is possible to arrange that as oxygen is so attracted into the space between the poles of a magnet it displaces a lightly suspended solid body by an amount which is taken as a measure of the oxygen content of the gas. In another instrument the oxygen in the gas is attracted to pass over a hot wire cooling it to a degree which is a measure of the oxygen content of the gas. Oxygen content of the flue gas is probably the most satisfactory indication of the combustion efficiency. Each fuel gives its best results with a particular proportion of excess air and this leads to an optimum residual oxygen content in each case—about 3 or 4 per cent for coke oven gas, 5 per cent for fuel oil, 6 or 7 per cent for coal burned on grates and even more up to 9 or 10 per cent for pulverized fuel. If oxygen exceeds the desired value and burner settings are correct the difference may be due to leakage through the furnace walls. This can be checked by a tracer technique usually using radon. Any negative deviation from the expected concentration as the tracer comes through the furnace must be due to dilution by infiltered air.

Fluid Flow Measurement

The determination of the delivery rate of fuel and air is as important as the more familiar metering of electrical energy.

With either gases or liquids the methods are similar. Where a pump or blower is being used its rev/min multiplied by swept volume is a measure of the rate of flow at the pump but cannot account for leakages. A more satisfactory device is an orifice meter in which the fall in static pressure complementary to a rise in kinetic pressure at a constriction is determined and converted to volume rate of flow (Fig. 70(a)). Assuming pressure loss due to friction and expansion to be negligible in so short a distance:

$$\tfrac{1}{2}\rho v^2 + p = \text{const.} \tag{28.21}$$

where ρ is density of fluid, v is velocity and p its static pressure. Considering two positions 1 and 2 just before the orifice and just at the vena contracta (narrowest point in the stream),

$$\tfrac{1}{2}\rho v_2^2 - \tfrac{1}{2}\rho v_1^2 = p_1 - p_2 \tag{28.22}$$

i.e.

$$\frac{v_2^2 - v_1^2}{2} = \frac{p_1 - p_2}{\rho} \tag{28.23}$$

If the orifice diameter $\not> 0{\cdot}2D$:

$$v_1 \ll v_2$$

and

$$v_2^2 \simeq 2\,\frac{p_1 - p_2}{\rho} \tag{28.24}$$

Then a volume flow can be calculated assuming a value for the diameter of the vena contracta and assuming the fluid to be incompressible. The absolute static pressure at the vena contracta and the temperature are also required if mass flow rate is to be determined.

In fact, these assumptions are not readily entertained and orifices are either calibrated against some other means of measurement like pitot tube surveys or a standard design (B.S. 1042) is adopted for which calibration tables are published, valid under the installation conditions prescribed. These can be used down to 2 in. diameter pipes. The orifice must be fitted in a length of pipe which is straight for about nine diameters, five upstream and four downstream.

Orifice meters are cheap and simple to insert and operate and give reasonable accuracy, about ± 1 per cent. They cause appreciable loss of total pressure, however, due to eddying immediately downstream from the plate. This can be avoided by using a nozzle (Fig. 70(b)) or a venturi throat (Fig. 70(c)) in which these losses are very much reduced. Similar installation precautions must be taken, and these also are prescribed in B.S. 1042.

The pitot tube is an L-shaped probe containing two compartments. An inner tube comes to an open end which faces upstream into the stream of fluid. An outer tube encloses the inner one and holes in its side lie parallel to the stream. The pressure in the inner tube is then the total pressure—static plus dynamic heads—while the pressure in the annular space between them is the static head only. The two compartments are connected to the limbs of a U-tube which indicates the magnitude of the dynamic pressure head $\frac{1}{2}\rho v^2$ (see Fig. 70(d)). Hence the gas velocity at the position occupied by the probe can be calculated, if ρ, p and T are also known. The whole cross-section of the duct can be surveyed and the mass flow rate computed. The pitot tube survey is generally accepted as the primary standard method of measuring flow rate against which other types of instrument can be calibrated. They are available in standard designs and must be used in a standard manner to achieve maximum accuracy.

A very elegant flowmeter is the "rotameter" (Fig. 70(e)) in which a light float is supported on an upward stream of fluid in a slightly tapering tube at such a position that its weight mg is just balanced by the difference between the static downward pressure

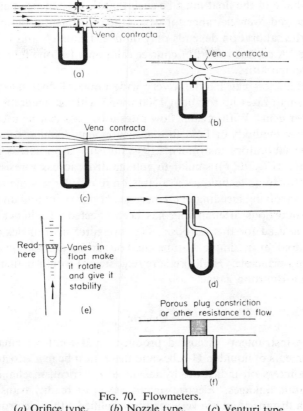

FIG. 70. Flowmeters.
(a) Orifice type, (b) Nozzle type, (c) Venturi type,
(d) Schematic Pitot tube, (e) Rotameter, (f) Resistance type.

on the top of the float (p_2) and the upward pressure, partly static (p_1) and partly kinetic ($\frac{1}{2}\rho v_1^2$). Where A is the cross-sectional area of the tube at the position taken up by the float and a that of the float the velocity of the fluid in the tube below the float is:

$$v_1 = \frac{2mg}{\rho a}\left(1 - \frac{a}{A}\right)$$

The shape of the float must be such that the turbulence pattern is unchanged over the range for which the instrument is designed so that the calibration depends only on a/A, i.e. on A, and is in fact linear. A calibration is, of course, valid only for one fluid and at one temperature.

Rotameters can handle a very wide range of flow rates, from very small rates up to about 1500 l./min with an accuracy about ± 2 per cent. With greater flow rates a by-pass can be used.

Other methods of fluid flow measurements include some very simple laboratory instruments, e.g. pressure drop across a constriction (Fig. 70(f)) similar to voltage drop across a resistance; and also some techniques available on a very large scale such as those involving bleeding in an inert gas at one point and analysing its proportions after mixing has been effected. Radioactive gas (Rd) is used for this purpose. Gas pressures in furnaces can be measured by a simple U-tube method or by adaptation of the aneroid principle. High pressure (e.g. steam) would be measured using a Bourdon gauge.

Control

The instruments discussed produce small e.m.f. or small displacements of liquid in U-tubes and these must be made to activate the pointers on indicators by means of electrical, mechanical or hydraulic linkages. Electric signals are most readily transmitted over a distance and are very readily amplified either electronically or by means of relays. Translation of the signal to action involves amplification to the point that valves may be opened and closed or switches operated. Electric motors or hydraulic pump systems are both employed in these operations. A study of servomechanisms and control systems is beyond the scope of this booklet, however, except for the following brief outline of some of the principles involved.

When a furnace is being controlled at a given temperature, ideally a steady state should exist in which the heat input exactly balances thermal losses in various directions including those to the cold charge. Any variations in thermal losses will upset this balance and will be detected as a change in furnace temperature. It is the function of the control system to adjust the heat input so that balance is again achieved, as quickly as possible, at the original temperature and with the new value of thermal requirement. In practice the thermal losses may vary continuously so that ideal control is never really achieved.

Most control systems are imperfect insofar as some degree of "hunting" develops—that is a cyclical variation of temperature is obtained about the desired value. This is due to several time lags which are inevitable in any control system. The most important of these is *Process Lag* which is inherent in the equipment, and in a furnace would be mainly due to the thermal inertia of the furnace and its load. This would be a complex function of mass, specific heat and conductivity. Secondly there is the *Measurement Lag* again unavoidable as the change in temperature cannot be detected before it happens and is more likely to be detected some time afterwards especially as many instruments make their measurements at specific intervals of time—say every 20 sec. Other causes include the temperature drop across the sheathing on the thermocouple. Thirdly, there is *Control Lag* which is a measure of the time taken for relays, switches, motors and valves to operate, and in some cases this could take several minutes.

There are several types of control system of varying degrees of complexity, used to achieve good control under different conditions.

(1) *Step Control* involves simply switching the input from one value to another. *Two Step Control* is simple "On/Off", or better "High/Low" switching, and *Multistep Control* might involve a series of steps such as "0, $\frac{1}{4}$, $\frac{1}{2}$, $\frac{3}{4}$, full". Insofar as none of the possible steps need correspond to the ideal heat input value, continuous hunting is almost inevitable, the time during which the input is "High" and "Low" respectively depending on the proportioning of the ideal value between the two possible values of heat input.

Step control is most successful where the capacity of the process (e.g. mass) is fairly high so that response to change of heat input rate is not too violent. The process lag should be low (i.e. good conductivity) and any changes in loading should be either small or slow as compensation does not become effective quickly. These control systems should have as little measurement and control lag as possible. If conditions are not as indicated some other control system should be used.

(2) *Floating* or *Integral Control* allows continuous variation of heat input rate between limits which may cover the whole range from "Off" to "Full On", but is more likely to cover only a part of the range. The result need not be very different from step control as the input can change to its extreme value and stay there till the sign of the deviation is reversed. The rate of change of input (speed of opening a fuel valve) may be constant or may depend on the magnitude of the deviation so allowing faster response to rapidly changing situations. Properly tuned floating control should be able to bring the fuel input rate to the ideal value for the required temperature in a few cycles of operations and maintain it even if the load fluctuates. It is particularly useful where the capacity of the process is low and it can accommodate large changes in load if these are not too fast.

(3) *Proportionate Control* sets the input to a value proportional to the magnitude of the deviation of the actual temperature from the required temperature and is usually set to operate in a band of temperatures, say, $\pm 50°C$ with respect to the desired value. Some other control is used to bring the temperature into that range from cold. This type of control can be used where the process lag is large provided the load changes are small. Input and temperature approach ideal values simultaneously from opposite directions.

(4) If the furnace load varies the ideal input appropriate to the required temperature will change and then *Proportional and Integral Control* should be used, in which the whole band "floats" so that the ideal input setting is changed upwards or downwards according to the success of the system in achieving its aim. A

furnace would require a lower input for a given temperature with its doors closed than if a big forging were sticking out of the door. Proportional and Integral Control could take care of a difference of this kind.

(5) *Derivative Control* is used when the controlled condition is fluctuating rapidly. The input rate is set at a value proportional to the rate at which the temperature deviation is altering. Thus some degree of anticipation is introduced. This control cannot be used alone but in conjunction with types 2, 3 or 4; it works best when the capacity of the process is high.

Obviously these systems involve complex instrumentation, much of it electronic in character and further developments incorporating computers are not unlikely. The secret of success in control is in the correct "tuning" of the control system to the characteristics of the unit being controlled, that is the timing of the checks and the magnitude of the changes in input rate effected on each occasion must be carefully chosen to suit the thermal capacity of the furnace and its load and the rate at which heat transfer can be effected within it.

Control may be applied to other functions of furnaces than straightforward heating. Pressure may be used to control dampers, gas analyses to control burner settings, or checker temperatures to effect reversals through regenerators. When it is fully understood how a furnace ought to run, and which operational factors are causes, and which effects, the next stage in development is to make the effects react on the causes in such a way that the furnace will carry on working in an ideal and thoroughly efficient manner.

Electronic Instrumentation

The greatest advances in instrumentation and control which have been made since about 1960, have been in the field of electronics. The rapid development of semi-conductor technology and the extensive replacement of vacuum valves by transistors has made the exploitation of small electrical signals by compact equipment both cheap and reliable. The same period has seen the rise of the digital computer as an indispensable tool in all branches of technology and the techniques of computing have quickly been applied to instrumentation.

The first stage in making a measurement or in exercising control is to produce an electrical signal by means of a "transducer". This simply means that as some particular characteristic of the system under observation changes, an electrical characteristic of the measuring system also changes. Some common transducers have already been mentioned and others are well known. A thermocouple in a furnace is a voltage transducer but thermistors and platinum resistance thermometers are resistance transducers. Simple devices like mercury expanding along a capillary tube to make contact with a platinum wire or a bimetallic strip acting upon a micro-switch can be considered to be resistance transducers in so far as they can change the resistance of a part of an electrical circuit from a finite value to infinity by breaking the circuit. Similar movement can be used to alter the inductance or the capacitance of a part of a circuit continuously. Pressure changes can be made to move a diaphragm between two plates in an a.c. circuit so altering its capacitance or the movement of the tube in a Bourdon gauge could shift an iron core in an induction coil so altering its reactance. The possibilities are endless.

When appropriate, the changes in resistance or reactance can be converted into a small current or voltage using an unbalanced Wheatstone bridge circuit in the case of d.c. signals or a suitably modified a.c. bridge for a.c. signals. It is sometimes possible at this stage to incorporate compensation for, say, temperature variations in the measuring system by opposing the measuring arm with a resistance which is subject to the same random fluctuations of temperature as is the measuring arm itself. The careful selection of bridge ratios can be used to effect some amplification of the original signal.

The amplification of electrical signals has traditionally been effected using circuits of greater or lesser complexity using thermionic vacuum tubes or valves. Today this would usually be done using transistors which serve the same purpose as the valves but are smaller, usually cheaper, and longer lasting. Individually, transistors are usually simpler than valves but used in groups they can perform as wide a range of functions and use much less power as they work. They require very little heating-up time and instruments need less cooling than those designed using valves.

Essentially a transistor is a tiny sandwich in which all three layers are made of either silicon or germanium of high purity except that the two outer layers have been deliberately contaminated (doped) with traces of one element while the inner layer has been doped with another element. Silicon has an effective valency of four and the two doping elements should have valencies of three and five to produce electron deficiency or excess in the material. Clearly there are two kinds of sandwich depending on whether the "meat" has an excess or a deficiency of electrons. Electrical connections are made to all three layers and in a simple application the current which can be made to pass through the sandwich will depend upon the "bias" applied to the middle layer. It thus behaves like a simple valve in which the current is controlled by the bias on the grid. If the signal from the transducer is applied to the "base" as the middle layer is called and a potential is applied across the outer layers the current produced will be (within certain operational limits) proportional to the signal and of much higher value (depending upon the potential used). If a single stage of amplification is not sufficient, further stages can be added and in general there is a degree of complexity built into amplifier circuits designed to obtain greater accuracy and higher gain without distortion of the original signal.

The amplified signal may be used to produce a dial reading on a galvanometer type instrument or to operate an electro-mechanical relay with direct effect on the system under control. There is a strong trend toward the use of digital read-out instruments and non-mechanical switching using transistor-like devices—silicon controlled rectifiers and switches. These can be included in the furnace heater circuit in series with the heating element and will pass a current only when the bias supplied by the control circuit exceeds a particular value. There is also the possibility of employing some degree of computation to the control—combining the information from one measuring device with that from others before signalling what action is to be taken. For this purpose as well as for digital read out, it is necessary to convert the amplified signal into a digital form. First it is converted by some "chopping" device to a square form a.c. signal which may be further modified. This must then be converted to a number—in binary notation— using a counting device. One way of doing this is to supply the signal current to a capacitor for a precise specified interval of time during which the capacitor will charge up and discharge to the counter many times. The counting would be done in binary notation using a bank of transistorized devices called "flip-flops". These tiny circuits have two stable states designated OFF and ON or ONE and TWO and a bank of, say, twenty of them would be able to spell out any number up to 2^{20} by virtue of the order in which they appeared in the ON or OFF state. For digital readout these would be converted to decimal but for computations they can be stored and operated upon mathematically. For example if the temperature of a furnace were measured at intervals the rate of rise or fall of temperature could readily be computed continuously and the rate of supply of energy could be adjusted according to a programme which might take into account simultaneously the difference between the true temperature and the control temperature, the rate of temperature rise, ambient temperature, the load in the furnace and whether or not the door was open. This could be controlled so that as the desired temperature was reached the rate of temperature rise would fall to zero.

There is no limit to the complexity which might be built into a control system but there must be economic justification for the cost involved. Large scale processes such as iron-making in blast furnaces and steelmaking in oxygen converters are currently operated under sophisticated computer control which embraces not only the energy input but the burdening also, taking into account the analyses of ores, fluxes and fuels as well as the temperature and composition desired in the product. The future will probably see an extension of this quality of control toward operations on a smaller scale as the cost of the equipment becomes relatively lower and the benefits which can be gained become better appreciated. The importance of the transistor in all this is that it makes possible an extremely large number of operations in a very short time. Nothing is moving except electrons, and these over very short distances, so that flip-flops can count many millions per second and a silicon controlled rectifier could switch a current on and off and on again, in three successive cycles of mains frequency. The systems outlined so briefly require a vast number of operations to be carried out. Transistor technology has made it possible to carry them out in almost no time at all.

Appendix

UNITS AND CONVERSION FACTORS

	Imperial units	c.g.s. units	S.I. units	Conversion factors
Length	in.	mm	mm	1 in. $= 2 \cdot 54$ cm $= 2 \cdot 54 \times 10$ mm
	ft	cm	m	1 ft $ = 3 \cdot 048 \times 10^{-1}$ m
		m		$ = 3 \cdot 048 \times 10$ cm
Mass	lb	g	g	1 lb $ = 4 \cdot 536 \times 10^2$ g
				$ = 4 \cdot 536 \times 10^{-1}$ kg
	ton	kg	kg	1 ton $= 1 \cdot 016$ tonne
				$ = 1 \cdot 016$ t
	short ton $(=2000$ lb)	tonne $(=1000$ kg)	(t = tonne)	
Force	lbf	kgf $(=kp)$	N	1 lbf $= 4 \cdot 536 \times 10^{-1}$ kgf
				$ = 4 \cdot 448$ N
				1 kgf $= 9 \cdot 807$ N
Energy	Btu	cal	J	1 Btu $= 5 \cdot 556 \times 10^{-1}$ CHU
				$ = 2 \cdot 52 \times 10^2$ cal
				$ = 1 \cdot 055 \times 10^3$ J
	CHU	kcal	kJ	1 CHU $= 4 \cdot 536 \times 10^2$ cal
				$ = 1 \cdot 899 \times 10^3$ J
	Therm $(=10^5$ Btu)	kWh		1 cal $ = 4 \cdot 186$ J
				$ = 1 \cdot 163 \times 10^{-6}$ kWh

	Imperial units	c.g.s. units	S.I. units	Conversion factors
Calorific value	Btu/lb	cal/g	kJ/kg	$1 \text{ Btu/lb} = 5 \cdot 556 \times 10^{-1} \text{ cal/g}$ $= 2 \cdot 326 \text{ kJ/kg}$
	Btu/ft^3	$(= \text{kcal/kg})$		$1 \text{ Btu/ft}^3 = 8 \cdot 9 \times 10^{-3} \text{ cal/l}$ $= 3 \cdot 725 \times 10 \text{ kJ/m}^3$
		cal/l	kJ/m^3	$1 \text{ cal/g} = 4 \cdot 186 \text{ kJ/kg}$
		$(= \text{kcal/m}^3)$		$1 \text{ cal/l} = 4 \cdot 186 \text{ kJ/m}^3$
Pressure	lbf/in.^{2}	kgf/cm^2	N/m^2	$1 \text{ lbf/in.}^2 = 7 \cdot 03 \times 10^{-2} \text{ kgf/cm}^2$ $= 6 \cdot 895 \times 10^{-2} \text{ bar}$
	lbf/ft^2	—	kN/m^2	$1 \text{ kgf/cm}^2 = 9 \cdot 807 \times 10^{-1} \text{ bar}$
	in. H$_2$O	cm H$_2$O	m bar	$1 \text{ in. H}_2\text{O} = 2 \cdot 491 \text{ m bar}$
	in. Hg	mm Hg	bar	$1 \text{ mm Hg} = 1 \cdot 333 \text{ m bar}$
		atm	$(= 10^5 \text{N/m}^2)$	$1 \text{ atm} = 1 \cdot 013 \text{ bar}$
Heat flow intensity	Btu/ft^2 h	cal/cm^2 sec	W/m^2	$1 \text{ Btu/ft}^2 \text{ hr} =$ $7 \cdot 535 \times 10^{-5} \text{ cal/cm}^2\text{sec}$ $= 3 \cdot 155 \text{ W/m}^2$ $1 \text{ cal/cm}^2 \text{ sec} =$ $3 \cdot 231 \times 10^{-3} \text{ W/m}^2$
Thermal conductivity	Btu-in./ft^2 h°F	cal/cm sec°C	W/mK	$1 \text{ Btu-in./ft}^2 \text{ h°F} =$ $3 \cdot 445 \times 10^{-4} \text{ cal/cm sec °C}$ $= 1 \cdot 442 \times 10^{-1} \text{ W/mk}$ $1 \text{ cal/cm sec °C} =$ $4 \cdot 186 \times 10^2 \text{ W/mk}$

Bibliography

1. BRAME, J. S. and KING, J. G. *Fuel: Solid, Liquid and Gaseous*, Arnold, 5th ed., 1956.
2. CREMER, H. W. and WATKINS, S. B. (eds.). *Chemical Engineering Practice*, Vol. 10, Butterworth, 1960.
3. *The Efficient Use of Fuel*, H.M.S.O., 1958.
4. SPIERS, H. M., *Technical Data on Fuel*, British National Committee, World Power Conference, London, 6th ed., 1961.
5. MACRAE, J. C. *An Introduction to the Study of Fuel*, Elsevier, 1966.
6. HARKER, J. H. and ALLEN, D. A. *Fuel Science*, Oliver and Boyd, 1972.
7. DAVIES, C. *Calculations in Furnace Technology*, Pergamon, 1970.
8. LEONARD, J. W. and MITCHELL, D. R. (eds.). *Coal Preparation*, A.I.M.E., 1968.
9. LOWRY, H. H. (ed.). *The Chemistry of Coal Carbonization*, 2 Vols. National Research Council (U.S.A.) 1945 and Supplementary Volume, Wiley, 1963.
10. TIRATSOO, E. N. *Natural Gas*, Scientific Press, 1972.
11. HOBSON, G. D. and POHL, W. (eds.). *Modern Petroleum Technology*, Applied Science Publishers/Institute of Petroleum, London, 1973.
12. *Modern Power Station Practice*, 8 Vols. Central Electricity Generating Board/Pergamon, London, 1971.
13. GLASSTONE, S. and SESONKE, A. *Nuclear Reactor Engineering*, Van Nostrand, 1963.
14. WEEDY, B. M. *Electric Power Systems*, Wiley, 1967.
15. THRING, M. W. *The Science of Flames and Furnaces*, Chapman & Hall, 2nd ed., 1962.
16. TRINKS, W. *Industrial Furnaces*, 2 Vols. Wiley, 4th ed., 1951.
17. ETHERINGTON, H. and ETHERINGTON, C. *Modern Furnace Technology*, Griffin, 3rd ed., 1961.
18. PASCHKIS, V. and PERRSON, J. *Industrial Electric Furnaces and Appliances*, Interscience, 1960.
19. FISHENDEN, M. and SAUNDERS, O. *An Introduction to Heat Transfer*, Clarendon Press, 1961.
20. McADAMS, W. H. *Heat Transmission*, McGraw Hill, 3rd ed., 1954.
21. CHESTERS, J. H. *Refractories: Production and Properties*, The Metals Society, London, 1973.
22. CHESTERS, J. H. *Refractories for Iron- and Steel-making*, The Metals Society, London, 1974.

23. LEVIN, E. M., ROBBINS, C. R. and MCMURDIE, H. F. *Phase Diagrams for Ceramists*, American Ceramic Society, 2nd ed., 1964.
24. JONES, E. B. *Instrument Technology*, Vol. 1, Butterworth, 3rd ed., 1974.
25. DIEFENDERFER, A. J. *Principles of Electronic Instrumentation*, Saunders & Co., 1972.

This selection may help students wishing to study the subject in greater depth than is possible with this small volume. The older books named may occasionally be found to be out of date with respect to details of practice but will generally be valid on points of principle. The good old general texts are not quickly replaced, modern authors apparently preferring to write monographs which frequently become dated very quickly.

(1), despite its age is still a useful textbook on fuels generally, though out of date on motor fuel, while (2) covers a similar field in a different style. (3) and (4) are reference books, the latter being an extensive tabulation of useful data. (5) and (6) are more modern textbooks than (1), very readable but much less comprehensive. (7) deals with combustion calculations, and problems in heat transfer and furnace aerodynamics at a fairly elementary level. (8)–(14) are monographs on coal (8), coking (9), gas (10), oil (11) and electricity (12), (13) and (14). (12) is in 8 volumes of which the 2nd on boilers, the 3rd on turbines, the 4th on generators and the 8th on nuclear power are likely to be the most useful. (13) is about nuclear power stations and (14) is about power transmission. (15) is a classic on furnace theory while (16), (17) and (18) are about furnace design and operation. (19) and (20) deal with heat transfer theory, (20) being the more advanced text. (21) and (22) together provide a most comprehensive treatise on refractories technology while (23) is an atlas of the appropriate phase diagrams. (24) deals with instrumentation in a traditional manner while (25) explains how electronics are being used in instrumentation and control.

Index

Abrasion resistance 250, 270
Absorption coefficient 165
After expansion 241, 270
Air, composition of 10
Alkalies 8, 262, 268, 277, 305
Alkaline earths 265, 305
Alkylation 63
Allotropy 273, 287
Alundum 232, 304
Al_2O_3-CaO-SiO_2 system 267
Alumina 302, 303
  bricks 271, 272
Aluminium 3. 301
  smelting furnace 226
Alumino-silicates 258–72
Amonia 35
Analysis, chemical
  of coal 18, 19
  of coke 19, 38
  of flue gas 9, 327
  of gaseous fuel 10, 12, 55
  of liquid fuel 9
  of refractories 251, 265, 288
  physical methods of 252
  proximate 11, 18, 38
  ultimate 10, 18, 38, 39
Andalusite 270
Anorthite 267–8
Anthracene 64
Anthracite 19, 21
Anti-knock 63, 72
Asbestos 237, 239, 301
Ash 7, 11, 12, 38, 39
  fly 8
  fusion point of 8
Asphaltenes 66
Atmospheres, controlled 138
Atomization, of oil 66

Bauxite 271
Benzene, benzole 64, 67, 71
Beryllia 305
Bitumen 61, 91
Bituminous coal 19
Black body 166
Bloating 243, 270
Boiler 226–30
  waste heat 49, 174
  water treatment 228
Boiling range of oils 61
Bonding
  chemical 256, 296
  glassy 277
  in basic bricks 283, 286
  mullite 245
  tar 256, 288
  trydimitic 245, 277
B.S. swelling number 20, 23
B.t.u., definition 13
Buoyancy 184, 185, 188
Burners 129–37
  Bunsen 132
  cyclone 128
  gas 133
  oil 131
  sonic 135
Butane 52
By-products 35

Calcium o-silicate 287
Calcium oxide 284
Calcium tri-silicate 256, 278, 287
Calcium tungstate 274
Calorie, definition of 13
Calorific value 13, 18, 40, 55, 67, 68
  net, gross 14, 15

CaO-SiO$_2$ system 278
Carbon 6, 11, 39, 67, 297-9, 302
  deposition 269
Carbonization 29-36
Castables 196
Cement, for bricks 238
Cetane number 70
Chamotte 254, 256
Charcoal 3
Chemical composition see Analysis
Chequers (checkers) 177, 270, 281
Chimneys 184
China clay 262
  bricks 270, 300
Chrome magnesite 292-6
Chromite 291-2
CHU, definition of 13
Ciment fondu 256
Clarain 21
Clay minerals 260
  plasticity of 262
Cleaning plant, gas 35-6
Clinker 8, 285, 287
Coal(s) 17-28, 98
  agglutination of 23, 32
  altered 17
  cannel 21
  classifications 18-24
  coking 19, 20, 29
  combustion of 125-9
  origins of 17
  pulverized 25, 27, 126-9
  rank of 19
  structure of 23
  storage of 26
  swelling of 22, 32
  use of 26-8, 90-2
Coalplex 28
Coal tar fuel 35-6, 64, 65
Cochrane abrasion test 41
Coke 18. 32-5, 37-44, 92, 121
  blast furnace 34, 37
  breeze 37
  combustibility of 39
  combustion of 121
  domestic 33, 37
  formed 33-4

oven 30-2, 35, 206, 280-1
  reactivity of 39
  strength of 40
Coking see Carbonization
Cold crushing strength 243, 294
Combustion mechanisms 121-38
Compensating leads 321
Composition see Analysis
Computers 338, 341
Conduction, conductivity, thermal
  150-6, 250, 301
  effective 171
  table of values 155
Connections, electrical 84
Control systems 234, 334-7
  hunting in 335
  tuning of 337
Convection 156-64
  forced 162-4, 224
  natural 156-61, 171, 225
Conversion factors 342
Converters 219-22
  Kaldo 221
  L-D 220, 256, 288
  rotor 221
Counterflow heating 175, 205-10,
  215
Cowper stove 178, 269-71, 281
Cracking, catalytic, thermal 60-3
Creosote 64
Cristobalite 259, 270, 274-6, 281,
  304
Critical air blast test 39
Cupola 216

Density
  apparent 38, 43-4, 250, 279
  bulk 38, 44
  real 38, 43-4, 66, 250, 277, 291,
  293
Desulphurization
  of gas 7, 35-6
  of oil 7, 48, 63
Devitrification 276, 304
Dickite 260
Dirt bands 7

Dissociation, of CO_2, H_2O 105
Distillation test 11, 65
Distribution of electricity 86–8
Doloma 288
Dolomite 286–90
Drying, of bricks 254
Dry pressing, of bricks 254, 264, 270, 283, 293
Durain 21–2
Dwight-Lloyd sintering machine 222–4

Economics 90–101
Eddying 183
Efficiency
  of conversion 85, 99
  of utilization 96–7, 172–83
Electric furnaces see Furnaces
Electricity
  consumption of 92–6
  distribution of 86–8
  production of 83–5
  use of 88–9
Electrolytic cell 226
Emissive power
  emissivity 165–6, 325–8
  of gases 168
Energy
  consumption of 90–6
  geothermal 5, 101
  hydroelectric 75, 92, 96
  nuclear 76–82, 92, 96
  reserves of 99–101
  solar 5, 100
  sources of 74
  tidal 5, 100–1
Excess air, oxygen 110, 330
Exinite 22
Expansion, thermal 193, 241–3

Fans 183, 189
FeO-SiO_2 system 248
Felspars 260
Fettling materials 238, 286, 290
Firebrick

analysis of 265
bloating 243, 270
classification of 264
high alumina 268–70
manufacture of 260–4
porous 300
properties of 266
siliceous 265
uses of 299
Fireclay 253
Firing 254, 264, 269, 277
Flame(s)
  diffusion 123
  luminous 64, 69, 170
  reactions 137
  temperature calculations 107–13
  temperature measurement 327
  velocity of propagation of 54
Flash point 11, 65, 67
Flint 275
Flip-flop 340
Flow of gases 183–92
Flowmeters 332–4
Forsterite 290–3
Free energy 105–6
Friction factor 186
Fuel(s)
  bed 122
  cell 82
  classification of 3
  consumption of 92
  gaseous 45–59
  liquid 60–73
  nuclear 76–83
  pulverized 4, 25, 64, 126–9
  smokeless 33–4
Furnaces
  air 201
  billet heating 208, 270
  blast 214–16, 218, 270, 298
  charging/discharging of 195
  classification of 197
  construction of 193–6
  crucible 197–200
  electric
    arc 88, 211–13, 290, 296
    capacitance 149

induction 88, 199–200, 232, 291
low shaft 217
resistance 88, 231
forced circulation 171, 224–5
foundations of 194
glass tank 204
hearth 200–13, 280, 287, 294
rotating 213
laboratory 231–6, 299
reverberatory 201–2
shaft 213–17
soaking pit 223, 292, 296
Wedge 208
Furnacite 33
Fusain, fusinite 22

Ganister 273
Gas(es) 30, 45–59
analyses of 10, 12, 55
Board 45
bottled 52
carbonization, from 30, 34–6, 52
classification of 45–6
cleaning of 35
coke oven 35, 52
combustion of 132–7
flow of 183–92
natural 45–8, 92
oil 51
preheating of 58
producer 48–9
production of 46–54
properties of 54–9
refinery 52, 91
reformed 52
retort 52
Town's 53
use of 58, 92, 96, 98
water 49–51
Gasoline 61, 67, 71, 91
aviation 73, 91
Ghelenite 267
Gibbsite 261, 303
Glass tanks 202–4, 271
Glass wool 301
Grashof number 158

Grates, moving 27, 223
Gray-King Index 20, 22
Grog 254, 269
Growth of bricks 294
Gunning 256

Heat
available 113–14
balance 179–82
by stages 112–13
diagrammatic 180
exchangers 175–9
of combustion 15
potential 109
sensible 109
transfer 150–71
waste 120, 174–6
Heating
capacitance 139, 149
electric arc 139, 143
induction 139, 144–9, 231
resistance 139, 231
Humic acid 6
Hydrogen 6, 30, 82
Hydrogenation 60, 65

Ignitability see Flash point or
Critical air blast or
Inflammability limits
Inflammability limits 54–5
Instrumentation 311–41
electronic 337–41
Insulation 173, 194, 233, 300–2
Iron spots 269

Joule, definition of 13

Kaolinite 260, 301
Kerosine 62, 91
Kieselguhr 301
Kiln
beehive 206
brick 206–7

cement 209, 296
Gjer's 217
rotating 209
tunnel 207
Kirchoff's Law 165
Knocking 70
Kyanite 270

Laboratory ware 299, 303-8
Ladles 195, 270
Lambert's Law 164
Laminar flow 157
Lance, fuel 58, 135
Le Chatelier Rule 57
Lignite 3, 18, 60
Liquid fuels
classification of 60
production of 60-5
properties of 67
tests for 65-6
uses for 68-9
Liquefied natural gas (lng) 47, 94
Luminosity 69, 127, 170, 213
Lurgi process, gas 53

Magnesia 302, 305
fused 303
sea water 284, 294
Magnesite 284
brick 284-6
properties of 289
-chrome brick 295-6
dead burned 284
Metakoalinite 264
Metal case bricks 286, 295
MgO-SiO₂ system 277, 279
Micum test 38, 41
Mineralizer 274, 291, 304
Mineral jelly 62
MnO-SiO₂ system 277
Mobility of clay 262
Moderator 78-80
Moisture 7, 11, 38, 66, 284, 287
Monotectic 277
Moulding of bricks 254 and see

Dry press
Muffles 88, 138, 178, 233, 237
Mullite 254, 258, 268, 270, 272
Mullite bricks 271, 304

Nacrite 260
Naphtha 62, 91, 94
Naphthalene 6, 35, 61, 64
Neutron 77
Nitrogen 7, 30, 67
Nuclear energy 76-82, 92, 95, 96,
100
Nuclear reactor 78-82
Nusselt number 158

Octane number 61, 72
Oil
bunker 91
burners 131
classification of 60-2
combustion of 129-32
diesel 67, 70-1, 91
fuel, heavy, light 62, 67, 91-3
gas 62
lubricating 62, 91
paraffin 62
petroleum, crude 60-1, 91
properties of 67
raw material, as 68, 91
refining of 61-4
sands and shales 60, 64, 100
signal 62
sources of 60, 91
tests on 65
transportation of 4
vapourizing 73
uses of 68-9
Olefines 61
Olivine 290
Orifice meter 332
Ovens 30-2 and see Coke
Oxygen enrichment 108, 115, 127,
130, 137, 214, 216, 221
Oxygen pressure 105-6
Oxygen in fuel 7, 30

Paraffin   6, 61–2
Paraffin wax   61–2, 91
PCE temperature   242, 316
Peat   3, 18
Peltier coefficient, effect   316
Pensky-Martin test   65
Periclase   284, 287
Permeability, of brick   250, 279
Petrography   252
Petrol   see Gasoline
Petroleum   see Oil
Picrochromite   291
Pinking   71
Pitch   64
   pulverized   65
Pitot tube   332
Plasticity of clay   262
Plumbago   298
Plutonium   79
Polymerization   63
Porcelain   304
Porosity
   of bricks   249, 279, 293
   of coke   44
Ports (gas)   133, 136–7, 203
Power
   nuclear   76–81
   tidal   5
   water   5, 75, 100
   wind   5, 100
Power factor   87, 144
Power sources   74–88, 100
Propane   52
Prandtl's number   158
Preheat
   of air   109, 119, 176–8
   of coking coal   32
Pressure losses   183–90
Pumped storage   75
Pyrometers
   calibration of   322
   contact   323
   optical   326
   radiation   324
   suction   323
Pyrometric chalk   316
Pyrometric cones   242, 315

Pyrometric fixed points   313–14
Pyrometric paint   316
Pyrometry   311–29

Quartz   264, 274–6
Quartzite   273
Quenching, of coke   32

Radiation   150, 164–71
   from flames   170
   from gases   167
   from solids   164
Reactivity, of coke   39, 122
Reactors, nuclear   78–82
Recuperators   174–7
Redwood viscometers   65
Reflectivity of coal   22
Reforming   63
Refractories
   classification of   237–9
   manufacture of   253–7
   properties of   240–52
   special   239, 257
   uses of   232
Refractoriness   242, 265, 271, 283,
      289, 290, 293, 294
   under load   243–6, 277–9, 283,
      289–91, 294
Regenerators   177–8
Reserves of energy   99–101
Resistors, electrical   140, 231, 298
   failure of   143
Residues from oil   62, 64
Retort   30, 217
Reynold's number   162, 185
Rotameter   332

Salt bath   225
Sand   273
Sand brick   256, 282
Sankey diagram   180
Seebeck effect   316
Seger cones   11, 242, 315
Semiconductors   337
Semi-silica   265, 282
Serpentine   290

Shale 7, 8, 17, 22, 24
  oil *see* Oil
Shatter test 38, 40, 42–3
Shock resistance *see* Spalling
Silcrete 273, 277
Silica
  in ash 7
  brick 237, 273–83
  high duty 273, 280
  vitreous 304
Silicon carbide 306
Sillimanite 270
Sintering 57, 222, 307
SiO$_2$-TiO system 277
Slag
  blast furnace 267
  resistance 247–9
  wool 301
Slip casting 254, 307
Smokeless fuel 29, 33, 34, 92
Soaking pit 223, 292, 296
Sodium sulphate 8
Sodium vanadate 8
Solvent extraction 22
Sosman's classification 274
Spalling
  structural 280, 283, 294
  thermal 264, 265, 271, 281, 283, 305
Special refractories 303–8
Specific gravity 11, 16 *and see* Density
Spinel 291, 303
Spirit, solvent, white 62, 91
Stefan's Law 164
Stirring, electromagnetic 195
Storage
  of coal 26
  of dolomite 287
Stream line flow 157, 186
Sulphur in fuel 7, 11, 30, 38–9, 66–7
Surface roughness 186
Swelling number 22–3

Tar 30
Temperature scales 312
Thermal cracking 62

Thermal efficiency 96–7, 172–83
Thermal expansion, reversible 241, 243, 265, 279
Thermistors 315
Thermocouples 316–23
Thermometers
  gas 314
  liquid 314
  resistance 315
Thomson coefficient, effect 316
Thoria 305
Tie-bars 193
Titanium dioxide 262, 265
Transducers 338
Transistors 339
Tridymite 259, 270, 274–6
Turbulence 157, 183–92

Units 13, 342
Uranium 76–81

Vaseline 62
Velocity heads 187
Venturi throat 187, 332
Vermiculite 301
Virtue diagrams 115, 117, 119, 175, 181
Virtue of energy 114, 172
Viscosity of oil 11, 65, 67
Vitrain, vitrinite 21, 22
Vitrification 264, 269
Volatile matter 11, 39
Voltage, line, phase, RMS 84
Volume changes 273, 287

Washing of oil 63
Waste heat 174
Waste heat boiler 49, 174
Water in fuel 7
Water cooling 49, 193–4, 214
Waxes 62
Wien's Law 167, 326

X-ray diffraction 252

Zebra roof 281
Zircon 305
Zirconia 283, 302, 305